A History of Three-Dimensional Cinema

A History of Three-Dimensional Cinema

David A. Cook

ANTHEM PRESS

Anthem Press
An imprint of Wimbledon Publishing Company
www.anthempress.com

This edition first published in UK and USA 2021
by ANTHEM PRESS
75–76 Blackfriars Road, London SE1 8HA, UK
or PO Box 9779, London SW19 7ZG, UK
and
244 Madison Ave #116, New York, NY 10016, USA

British Library Cataloguing-in-Publication Data
A catalogue record for this book is available from the British Library.

Library of Congress Control Number: 2021942101

ISBN-13: 978-1-83998-012-1 (Hbk)
ISBN-10: 1-83998-012-5 (Hbk)

Cover image: By Sashkin/Shutterstock.com

This title is also available as an e-book.

For my grandchildren (in their birth order), Allison, Rebecca, Sydney, and Colin—the future of 3D and everything else.

CONTENTS

FIGURES

ACKNOWLEDGEMENTS

First, I want to thank Dana Polan who read the manuscript in its early stages and let me know that I was headed the right direction. Timothy Corrigan and Bruce Kawin read the completed manuscript for Anthem Press, offering many valuable and detailed suggestions that helped to make this a better book, and I owe them a great debt of thanks too.

At the University of North Carolina Greensboro, I need to thank Vice Chancellor of Research and Engagement Terri L. Shelton, Dean of the College of Arts and Sciences John Kiss, and Department of Media Studies Head Kimberlianne Podlas, who helped substantially in gaining the grants and research leave that made this book possible. Thanks to my colleagues in Media Studies who read and commented on my original research proposal—Jeff Adams, Matt Barr, and Michael Frierson, the last of whom also helped me to work out some of the technical issues involved in my illustration program. In addition, Brett Ingram, former colleague and independent author/filmmaker in Greensboro, provided helpful feedback for this project, as did Adjunct Professor of Media Studies (and my co-editor for the Anthem Series on Exploitation and Industry in World Cinema) Will Dodson.

I would also like to thank my editor Megan Greiving and the team at Anthem Press for their invaluable insight, assistance, and patience during the making of this book.

Finally, tremendous thanks must go to my wife, Diane Holt Cook, who has been my steadfast partner in life, in work, and all things for more than 50 years.

PREFATORY NOTE/INTRODUCTION

Most of the three-dimensional (3D) cinema discussed or mentioned in this book originated in the United States and Europe, with some nods to work done in China, India, and Japan. Many other countries experimented with stereoscopic motion pictures in the early 1950s (e.g., Hungary, the Soviet Union, West Germany, Italy, and Mexico), but each produced only a few films in the process and experienced nothing like the 1952–55 boomlet in Hollywood. In the post-*Avatar* period, 2010–20, digital 3D movies were widely popular and produced by many non-Western industries, but only a handful of such films appeared outside of their domestic markets either in theaters or on 3D Blu-ray discs. When the COVID-19 pandemic of 2020–21 ravaged the world's film industries, stereoscopic films nearly disappeared altogether because they could not be viably exploited in either theaters, due to the possibility of viral infection, or homes, due to the scarcity of 3D-capable televisions since 2016. The omission of non-Western cinema cultures from this study is therefore due to a lack of accessibility rather than critical choice.

Chapter 1

"A NEW WAY TO SIMULATE PRESENCE": THE FOUNDATIONS OF STEREOSCOPIC ENTERTAINMENT, 1427–1888

Stereoscopic prehistory

Most of us see the world in three dimensions. This is so primarily because humans possess binocular vision, although other factors such as foreground overlap, gradient texture, and relative size are important to depth perception too. (For example, the illusion of depth on the flat surfaces of paintings, drawings, and photographs depends on the replication of these alternate depth cues through the system of linear perspective.) Because our two eyes are aligned so that their fields of vision partially overlap, causing a slight parallax in which each eye sees objects from a slightly different point of view, their combination in our minds creates binocular vision. The distance between the two eyes of a normal adult is almost invariably 6.5 cm, and this binocular disparity provides cues that the brain automatically interprets as depth.

Humans since ancient times have also been understood as beings who create and manipulate signs—things that take the place of or stand for other things. We are the "representational animal," or *animal symbolicum* in the term of early twentieth-century philosopher Ernst Cassirer. Representation (or "re-presentation"), significantly, is the central characteristic of art. So, it is not unnatural that three-dimensional representation would become an important issue for aesthetics early in our history. By the middle Renaissance, such concerns were manifest in the development of linear perspective in painting and drawing, as noted above; the disposition of scenographic space in the theater; and the manipulation of spatial perspective in landscape architecture, sculpture, and bas-relief. Contemporaneously, at the level of popular entertainment, a cultural precedent for three-dimensional cinema appeared in the form of the European peepshow, which in the eighteenth century spread to the United States, China, and Japan.

According to Richard Balzer, a peepshow is a closed or semi-closed box having at least one hole through which a perspective is seen, which may or may not use mirrors to create its illusion; the factor of enclosure is important, because it gives the viewer the sense of approaching an inaccessible space and seeing a hidden image, similar to the voyeuristic experience of watching a movie in a darkened theater. Its origin seems to date from the work of Italian painter and architect Leone Battista Alberti, who in 1437 developed an instrument to view perspectives through a small hole in a box. Balzer points out that Alberti's device appeared at a time when many artists were trying to understand the phenomenon of linear perspective, employing a central vanishing point in their work "to transform two dimensional art into the appearance of three dimensional reality."[1] (Not coincidentally, around 1450, Alberti designed the Tempio Malatestiano in Rimini, a pioneering classicist building that uses bas-reliefs by Agostino di Duccio inside and on its external walls.) The Italian physician Giovanni Battista della Porta popularized the *camera obscura* ("dark room") in his treatise *Magia Naturalis* in 1558, and a group of Dutch painters began to use "perspective boxes" at about the same time. These created the illusion of objects occupying space using mirrors and painted interior surfaces; the point of perspective appeared when the viewer, presumably a painter attempting to create linear perspective on his canvas, looked through a hole in the box. Perspective boxes, known as "peeping machines," eventually found their way to Japan and China where Western-style perspective began to appear in paintings as early as 1683.

By the dawn of the eighteenth century, the peepshow had shed its scientific mantle and become fairground entertainment. Balzer notes that there were actually two types of peepshows, one of which exploited depth. These devices included a lens in their viewing holes to focus on framed images located farther back in the box, which could be backlit by sunlight or a candle; they were also capable of displaying multiple views, some using hand-cranked rollers to make the change.[2] The other type of peepshow, called the *boite d'optique*, used a viewing lens in the peephole and a mirror placed at a 45-degree angle that redirected the eye toward the view; these instruments might use proscenium frames but had no mechanisms for changing views. The widespread popularity of peepshows led to the large-scale production of views between 1740 and 1840 in several major urban centers, including London, Paris, Augsburg, Amsterdam, and Vienna. The images were made in a standard size from copper plate engravings, with most printed on thin paper, and some had pin-pricked perforations that allowed them to be transformed through backlighting (e.g., windows in buildings could twinkle in the dark,

1 Richard Balzer, *Peepshows: A Visual History* (New York: Harry N. Abrams, 1998), 18.
2 Ibid., 28.

daylight scenes could reappear at night, etc.). By the late 1700s, peepshows had begun to enter the home, and in the 1820s engravers started to create small, accordion-style peepshows as souvenirs to commemorate historical events and celebrations.[3] These were equipped with paper bellows that could be easily contracted or expanded, so that the viewer could look through a tiny hole and encounter layer upon layer of images, like a pocket-sized stage set. In the late 1890s, peepshows were largely replaced by motion pictures, although they provided the model for Edison's coin-operated Kinetoscope, the first widely marketed motion picture viewing device, as well for the Dickson and Casler Mutoscope, which used flip book technology to display moving photographic images to one person at a time. In the early twentieth century, such peepshows became the province of erotica at penny-arcades and, during the 1980s, XXX-rated pornographic videos in the "pay-and-spray" booths of adult bookstores.

Three-dimensional entertainment: The optical intervention

The word "stereoscopy" is derived from Greek, where *skopion* is "to see," and *stereo* is "solid": stereoscopy, in other words, means to see the solidity of things—their three-dimensionality, that is—as most normally functioning humans do in real life. The term was not coined until 1833, but the principle on which it is based was first articulated by the mathematician Euclid in ancient Greece. In his *Treatise on Optics* (c. 300 BC), he noted that the left and right eyes see different, geometrically measurable views of a sphere. Later, in the second century AD, the Greek physician Galen observed the same phenomenon and wondered how it is that our eyes resolve two separate images into one, conjecturing that the singleness of vision is explained by a crossing of fibers in the optic nerve. (Some of these fibers actually do cross, but that is not the reason we see two images as one.) In the sixteenth century, Giovanni Battista della Porta cited both Euclid and Galen in his *On Refraction* (1593) to explain why we don't see double images, theorizing that vision consists of seeing through one eye and then the other in rapid succession. In 1611, German astronomer Johannes Kepler offered a projection theory of vision in which he imagined mental rays traveling outward from the eyes in straight lines, which would construct a single image at the point of their intersection. These and similar attempts to explain how two-eyed vision becomes one led finally to the invention of the stereoscope by the English scientist Sir Charles Wheatstone (1802–1875) in 1833. This was a large refracting mirror device for the viewing of simple paired line drawings, and later paintings, in depth. Wheatstone

3 Ibid., 32; 38.

subsequently stated in his *Contributions to the Physiology of Vision* (1838) that in human binocular vision, the lenses of our eyes project two slightly different images onto the retinas, and our brain calculates the distance between them as depth. Stereoscopy replicates this process by providing left-eye views and right-eye views (stereo pairs) of the same picture at slightly different angles, which when viewed simultaneously creates the illusion of depth (stereopsis). Over time, many viable systems for the production and exhibition of three-dimensional images have been innovated in both the analog and digital realm. All of them, however, rely on the same optical principles of Wheatstone's stereoscopic line drawings nearly two centuries ago.

Once it was codified, there were initially two different ways to experience stereopsis in the nineteenth century—the stereoscope for private viewing and the stereopticon for mass presentation.

With the advent of daguerreotypey in 1839, drawn and painted pairs were replaced by photographic ones, and in 1844—following the principles articulated by Wheatstone—Sir David Brewster (1781–1868) invented a handheld apparatus for the individual viewing of stereoscopic images through a system of prismatic lenses, with the stereo pairs mounted on a single card (the ancestor of the ever-popular View-Master, introduced to the mass market in 1939). Brewster, already famous for his invention of the kaleidoscope in 1816, used magnifying lenses to shorten the distance between the viewer's eyes and the stereo pair, and in 1856 he published *The Stereoscope: Its History, Theory, and Construction* (London: John Murray), claiming that he, not Wheatstone, had discovered binocular perception. Despite this arrogant deception, Brewster made important contributions to the practice of stereoscopic photography, establishing that the camera lenses should have the same aperture as the human eye and that the viewer's lenses should approximate the average separation between the human eyes (as stated earlier, 6.5 cm, or about 2 ½ inches).

The leap into photography came in the spring of 1850 when Brewster demonstrated his stereoscope in Paris and several eminent opticians—most prominently Jules Duboscq—formed an association to have a large number of them constructed for commercial sale. Their first images were provided by daguerreotypey, a photographic process introduced by Louis Jacques Mande Daguerre (1789–1851) in 1839 that fixed a detailed image obtained by a camera obscura on a silvered copper plate.

Before that occurred, however, Daguerre, in partnership with Charles Marie Bouton, had invented the diorama, which was first exhibited in Paris in 1822 and in London a year later, as a form of immersive public entertainment. The term originated from the Greek *di*, meaning "through," plus *orama*, meaning "that which is seen," or "a sight." The original diorama was a theater with a massive turntable on which audiences would stand to view a landscape that was hand-painted on both sides of a linen panel. When lit from the front the scene would be shown in one state and when lit from behind in another (daylight would change to moonlight, winter to summer, an event like it would be shown in before and after pictures, etc.); then the audience would rotate to view a second and sometimes a third panel in a show that lasted between 10 and 15 minutes. Paintings employed vanishing perspective to create the illusion of depth, and later dioramas added sound effects, prefiguring three-dimensional cinema. A successor to the diorama was the panorama (from the Greek "to see all") or cyclorama, in which a continuous 360-degree painting was displayed in a cylindrical space constructed so that the audience was surrounded by it; these were often devoted to scenes from famous battles, such as the Battle of Gettysburg (1863; panorama painted 1883) and the Siege of Paris (1870–71; panorama painted 1891), and some were later motorized so that the audience could remain stationary as the painting circled around it. (Occasionally, faux terrain and life-sized plaster figures would be placed in front of the painting to enhance its 3D illusion; during the Victorian era, panoramas in Europe and North America saw hundreds of thousands of visitors a year.) As the turn of the century approached, virtual voyages became popular in panoramas simulating balloon trips, exotic train rides, and ocean voyages, including one patented in 1895 as the "Stéréorama mouvant" ("moving stereorama") that represented a sea voyage along the Algerian coast. Perhaps the most spectacular of these was Raoul Grimoin-Sanson's Cinéorama, built for the Paris Exposition of 1900, which fitted a panoramic rotunda with 10 synchronized motion picture projectors arranged in a 360-degree arc around the viewing platform to simulate a balloon ride over the city.[4]

By mid-century, three-dimensional photographs were all the rage at public exhibitions such as London's 1851 Crystal Palace Exposition (where Queen Victoria herself examined Brewster's instrument and gave it her endorsement), and during the 1870s the stereoscope, and its accompanying collection of stereoscopic views, became a common household item in England and continental Europe. In 1861, poet and stereo enthusiast Oliver Wendell Holmes, who had over 1,000 stereo cards in his personal collection, designed a lightweight, streamlined version of the instrument that was patent-free and became wildly popular in the United States over the next 50 years. Later in the nineteenth century, free-standing tabletop viewers were invented that could present many slides in succession; these so-called "automatic viewing cabinets" could hold between 50 and 250 slides in a revolving belt mechanism, yet they were too expensive for mass consumption. By the turn of the century, however,

4 Errki Hutamo, *Illusions in Motion: Media Archaeology of the Moving Panorama and Related Spectacles* (Cambridge, MA: MIT Press, 2013), 306–7; 317–18.

handheld stereography had truly become a mass medium, with firms such as Underwood and Underwood in New York selling over 10 million cards a year that provided stereoscopic views of world travel sites, news events, portraiture, and dramatic fiction. (Much American and European scenery and many statues, ancient and modern, were recorded *only* for stereo photography because depth perspective was their most prominent feature; other popular subjects included American Indians, the Civil War,[5] architecture, and astronomy.) As if to confirm the ubiquity of stereoscopic photography, in 1901 Kodak brought out the No. 2 Stereo Kodak and the Stereo Brownie, simple cameras that extended the power of three-dimensional picture-taking to the common man.[6]

As with the invention of motion pictures several decades later, attempts to solve a scientific problem—in the case of movies, how to measure movement too rapid for perception by the human eye; in the case of stereoscopy, why our eyes see two separate images as one—resulted in the innovation of a new form of industrialized entertainment.

Stereoscopic image projection

On the projection front, in 1858 the French physicist J. Charles D'Almeida proposed two methods for screening stereoscopic images: (1) an additive-color system that projected two separate transparencies, or slides, simultaneously, and (2) a stroboscopic system in which two separate pictures would be flashed forth in rapid succession. Both methods required the use of two magic lanterns[7]

---

5 In fact, according to R. M. Hayes, much of the existing photographic documentation of the Civil War came in the form of stereoscopic slides (*3-D Movies*: 1). See, for example, Glenn Bonat's BDR compilation *Stereographs: 3D Portals to the Past—Episode One: Photographs from the American Civil War Era in 3-Dimensions* (Helicon Digital Arts, 2018).

6 Hal Morgan and Daniel Symmes, *Amazing 3-D* (Toronto: Little, Brown, 1982).

7 Invented in 1659 by the Dutch scientist Christiaan Huygens, the magic lantern (Latin: *laterna magica*) was the first projector. By focusing light from a lamp through a lens, it could project an image that had been painted on glass onto a larger surface. Since it seemed to make images appear out of nowhere, the word "magic" was attached to the lamp's name. Like so many other optical devices, the magic lantern was first developed in a scientific context, but it soon came to be used for the purpose of commercial entertainment. By the eighteenth century, the lantern had become the province of itinerant showmen who carried the instrument and a selection of glass slides on their backs from place to place, showing them to small paying audiences all over Europe and the Americas. In the 1820s, the invention of oxy-hydrogen limelight meant that lantern shows could be projected to large audiences; slides became more elaborate and brilliantly colored, and they began to be produced in large quantities. The development of photographic slides, known as "hyalotypes," by William and Frederick Langenheim in Philadelphia in 1849 paved the way for their mass production. At the same time, lanterns themselves became smaller, more portable, and available to the common man for use in

to superimpose the images at slightly different angles on-screen. In the first, images were colored red and green by filters on the projection lenses and viewed through goggles with corresponding left- and right-colored cels to produce a stereoscopic illusion. In the second, synchronized shutters would rotate in front of both the projection lenses and the eyes in the goggles at an exchange rate fast enough to produce stereopsis. (Almeida later suggested that the goggles' shutters could be operated electrically for viewing stereoscopic motion pictures, forecasting the use of active-shutter glasses for digital 3D cinema in our own time.) The first method, which became known as "anaglyphic" 3D, was the basis for a popular form of theatrical entertainment later in the century when duo-colored slides would be projected by a two-lens "bi-unal" magic lantern, called a stereopticon, onto a large screen for spectators wearing correspondingly colored glasses. The effect was to produce true three-dimensional images for a large audience on a big screen, in the same way that the stereoscope produced them for individuals at home. (Because the producers of these slides often cut the double images in half for "flat," or planar, projection through a single lens, the public came to call virtually any projector of photographic slides a "stereopticon.")[8] Furthermore, even when they were monoscopic, the high definition of photographic slides projected with bright oxy-hydrogen limelight made them *seem* three-dimensional to contemporary audiences, much like nonstereoscopic 4K OLED TV monitors do today. D'Almeida's second method became known as the "eclipse" technique, since the synchronized shutters caused the viewer to see only the image targeted for the appropriate eye and blocked out the one intended for the other in a quickly cadenced cycle.

All new technologies of representation are eventually, if not always initially, appropriated for purposes of pornography (e.g., VHS in the early 1980s, streaming video on the Internet since the late 1990s), so it should come as no surprise that images of the female nude were among the first widely circulated daguerreotypes. Of course, such nudity had been the subject of the legitimate arts of drawing and painting since the Renaissance, and because photography

such public venues as churches, meeting halls, and lodges, and even in the private home. According to Terry and Deborah Borton, by 1895 magic lantern shows had become a mass entertainment medium, and there were over three hundred manufacturers and distributors of lanterns and slides in the United States, the largest of which offered 150,000 slides for sale, many of them with mechanisms that added movement to the projected image. (*Before the Movies: American Magic-Lantern Entertainment and the Nation's First Great Screen Artist, Joseph Boggs Beale* [London: John Libbey, 2014]: 11–12.)

8 In fact, according to Denis Pellerin and Brian May, many such stereo "halves" are used to illustrate books about nineteenth-century history, even though they are not credited as such. (*The Poor Man's Picture Galley: Stereoscopy versus Painting in the Victorian Era* [London: London Stereoscopic Society, 2014]: 11.)

modeled its practices on these earlier arts, the representation of female nudity made perfect sense. So too with stereography. In 1856, D'Oliver of Paris published a large collection of stereoscopic nudes for sale to the public. According to Serge Nazarieff, they were printed on glossy paper, executed with great skill, and cleverly tinted. While D'Oliver's stereo sets were tastefully rendered and not pornography as such, it didn't take long for stereography to appropriate the sex act and sexual fetishes as subjects.[9] As Jonathan Crary writes in *Techniques of the Observer*:

> In devising the stereoscope, Wheatstone aimed to simulate the actual presence of a physical object or scene, not to discover a new way to exhibit a print or drawing. [...] It is no coincidence that the stereoscope became increasingly synonymous with erotic and pornographic imagery in the course of the nineteenth century. The very effects of tangibility that Wheatstone had sought from the beginning were quickly turned into a mass form of optical possession.[10]

A more elevated use of the stereoscope involved the recreation of scenes from well-known Victorian paintings with live models. During the 1850s, the appearance of a new painting by a distinguished artist was big news, but many could not afford to visit a gallery or salon to directly view the works. Thus, enterprising photographers created stereo cards for commercial sale that used real actors to reconstruct the scenes and gave the man in the street the opportunity to view them at home in 3D; these were known collectively as "the poor man's picture gallery."[11] Scenes from popular plays, operas, and *tableaux vivants* were similarly recreated in photographic studios using either live actors or clay figures with miniature scenery, the most notable of which was *Les Diableries* published in Paris during the 1860s. This stereoscopic series depicted scenes from the daily life in Hell in sculpted clay vignettes similar to modern slides, many of them satirizing the cultural decadence of Napoleon III's Second Empire.[12]

---

9 Nazarieff claims that there was a virtual explosion of both formal nude studies and pornography during the Second Empire, 1852–70, as stereo prints on albumen paper were replaced by prints on gelatin silver bromide paper and photo-engraved postcards—the notorious "French postcards" of popular lore. (Serge Nazarieff, *The Stereoscopic Nude, 1850–1930* [Benedikt Taschen Verlag Berlin, 1990], 15–16.)

10 Jonathan Crary, *The Techniques of the Observer: On Vision and Modernity in the Nineteenth Century* (Cambridge, MA: MIT Press, 1992), 122; 127.

11 Pellerin and May, *The Poor Man's Picture Gallery*, 12.

12 Brian May and The London Stereoscopic Society have produced a seven-minute, originally orchestrated animated compendium of stereo slides from *Les Diableries* as *One Night in Hell* (2014), which appears as a supplement on the 3D Film Archive Blu-ray

In *Stereoscopic Cinema and the Origins of 3-D Film*, Ray Zone points out that "the utopian dream of stereoscopic images in motion was very much alive from 1850–1875" and that many inventors, photographers, and optical specialists worked on methods to achieve it.[13] Among those was Johannes Czermak, a Bohemian who, in 1855, designed a form of vertical Zoetrope in which internal mirrors reflected stereo pairs that seemed to move through persistence of vision. In 1859, a French toymaker named Victor Pierre Sies patented an "Animated Stereoscope" that used hand-colored albumen prints to create moving stereo pairs in a similar device. Between 1859 and 1861, the Belgian civil engineer Henry DuMom patented 12 different devices, all called the "Omniscope," designed to put stereoscopic photographs into intermittent motion for viewing through anaglyphic glasses. It was Charles-Émile Reynaud (1844–1918), however, who created the nearest relative of three-dimensional motion pictures as a by-product of his "Théâtre Optique," patented in 1888. A refinement of his "Praxinoscopic Theater" of 1877, which animated drawings using a system of rotating drums and prismatic mirrors, the Théâtre Optique projected long, sprocketed bands of hand-painted transparencies onto a screen in animated entertainments known as *Pantomimes lumineuses*, some of which ran from 12 to 15 minutes. Between 1892 and 1900, Reynaud gave 12,800 performances of these shows at the Musée Grévin in Paris to over half a million visitors, and he eventually replaced his painted images with color-retouched photographic ones in order to compete (unsuccessfully, it turned out) with the newly invented technology of motion pictures. Early in the century, he began to work on a device for viewing short stereoscopic sequences called the "Stereocinema," which resembled a dual vertically mounted praxinoscope and was in fact patented in 1907 as the "Stereoscopic Binocular Praxinoscope." Only one of these was ever built, and it survives today, although Reynaud destroyed most of the rest of his work, including all but two of his *Pantomimes* bands, in a fit of depression shortly before his death.[14]

---

restoration of *The Mask* (Kino Lorber, 2015). This is an adjunct to the book by Denis Pellerin, Brian May, and Paula Fleming titled *Diableries: Stereoscopic Adventures in Hell* (London: London Stereoscopic, 2013).

13 Ray Zone, *Stereoscopic Cinema and the Origins of 3-D Film, 1838–1952* (Lexington: University Press of Kentucky, 2007), 28.

14 He threw it all into the Seine, together with a purpose-built stereoscopic camera and film footage of himself that he had taken with it. Zone, *Stereoscopic Cinema*, 28–29.

Chapter 2

"A VERY VIVID IMPRESSION OF MOVEMENT": EARLY 3D CINEMA, 1895–1952

Motion pictures

The source of Émile Reynaud's despondency was apparently his failure to have invented true motion pictures, in which he was preceded by the Edison Corporation in 1893 and Lumières Freres in 1895. The Edison Kinetograph, actually "invented" by laboratory assistant William Kennedy Laurie Dixon (1860–1935), was a clever synthesis of existing principles and techniques derived from Eadweard Muybridge (series photography), Étienne-Jules Marey (chronophotography), and others that used celluloid roll film in a battery-driven camera to take photographic exposures of things in motion at the approximate rate of 40 frames (and later 16 and 24) per second. The Kinetograph incorporated what have come to be recognized as the two essential components of motion picture camera engineering: (1) a stop-motion device to ensure the regular but intermittent movement of the film strip through the camera (adapted by Dixon from the escapement mechanism of a watch) and (2) a perforated film strip (first innovated by Reynaud for his Théâtre Optique). Intermittency permits the unexposed film strip in its transit through the camera to be stopped for a fraction of a second before the lens while the shutter opens to admit light bouncing off the subject and expose the individual frames. The synchronization of the film strip and the shutter is accomplished by the perforations in the film strip, which is pulled through the camera by a system of clawed gears known as sprockets. This process is reversed in the projector or viewing device, with the perforations—or sprocket holes—ensuring the exact regularity of the film strip's movement as it passes through both machines. Because it was battery-driven, the Kinetograph was virtually inert, weighing about 600 lbs, and it could only be used inside of a purpose-built studio called the "Black Maria." Furthermore, Edison did not initially commission the designing of a projector, so motion pictures made with the Kinetograph were first seen in a coin-operated peepshow device called the Kinetoscope. (In 1896, Edison bought the rights to a projector invented by C. Francis Jenkins and Thomas Armat and patented it as the "Edison Vitascope," so that Kinetograph films could then be exhibited to large audiences.) The Lumiere's Cinématographe adapted the engineering principles of Edison's machine to a much smaller casing and was hand-cranked, weighing approximately 16 lbs, so that it was easily portable and suitable for outdoor shooting. The Cinématographe could also be used as a film printer and projector, meaning that films shot with the device could be rapidly processed and shown to the public.

Early experiments with three-dimensional cinema

With the development of motion pictures, it was natural that the medium's pioneers would experiment with anaglyphic 3D systems, since the photographic principle was the same as for stereoscopic slides. If two strips of film, one tinted red and the other tinted green (or cyan—greenish-blue), were projected simultaneously for an audience wearing glasses with red- and green-tinted lenses, the effect was a stereoscopic synthesis in monochrome. In other words, the color of the eyepiece lens and the corresponding color of the projected frame were used to maintain binary vision. (Edison and Dickson anticipated stereoscopic motion pictures, or at least the 1893 Kinetograph patent application claimed as much ["the reproduction of stereoscopic photographs of moving objects gives a very vivid impression of movement"] and contained a schematic drawing of a twin-lensed camera, but this was seemingly never built.[15]) It is often claimed that William Friese-Green (1855–1921), a British inventor and professional photographer, built the first anaglyphic motion picture camera in 1889; however, the device was actually designed by Frederick Henry Varley and patented by him in 1890. Friese-Green used this instrument later that year to make a stereoscopic film of some family members entering Hyde Park, footage from which survives today. In 1893, he registered a patent for a camera almost identical to Varley's, but it operated at a speed too slow for successful stereoscopic projection.[16]

Nevertheless, it was the first motion picture camera to combine both left and right images of a stereo pair onto a single strip of film, and it probably influenced subsequent innovations. Friese-Green's later contributions to early

15 T. A. Edison, "Apparatus for Exhibiting Photographs of Moving Objects." U.S. Patent No. 493,426 (March 14, 1893): 3.

16 In 1893, Edison's assistant William Kennedy Laurie Dickson also filed an application for a stereoscopic motion picture camera (British Patent No. 6,794), based on an earlier Edison design. The camera was apparently never built.

film technology included building, with John Alfred Prestwick, a twin-lens projector intended to eliminate flicker in 1896 and patenting an early color film process similar to Charles Urban's Kinemacolor in 1905. His work is celebrated, somewhat fancifully, in the romantic biopic *The Magic Box* (John Boulting, 1951), produced for the Festival of Britain, and his tombstone bears the aspirational inscription "The Inventor of Kinematography."

Auguste and Louis Lumiere (1862–1954; 1864–1948), French photographic manufacturers who invented the Cinématographe in 1895, also experimented with anaglyphic 3D. One of their first films had been *L'Arrivée d'un Train en gare* (*Arrival of a Train at a Station*, 1895), a short *actualité* in which a locomotive enters the station at La Ciotat on the diagonal by moving insistently toward the camera lens. The realistic depth perspective on-screen supposedly frightened early audiences who thought the train was going to run them over. At the very least, it impressed them, so the Lumieres began to contemplate remaking the film in 3D. In 1900, Louis took out a patent for stereoscopic moving images based on a rotating octagonal disc, which was apparently intended for viewing through a Kinetoscope-like peepshow device. While this system may have been publicly demonstrated in 1903, it was not until 1935 that *L'Arrivée d'un Train en gare* was remade in three dimensions, using an apparatus with prisms and lenses that placed the left- and right-eye frames one on top of the other on horizontally traveling 35mm film.

In addition to the anaglyphic method, polarized light was another way to separate left and right images for stereoscopic projection during the late nineteenth century. Polarization is possible because light can be made to travel at different angles when projected through a prism, a phenomenon known since 1669, when a Danish physician named Erasmus Bartholinus noted the double refraction of light through an Iceland spar crystal. In 1828, the Scottish physician William Nichols split a similar crystal to transmit light in two directions, and in 1895 a British physician named John Anderson invented a stereoscopic projector that, according to his patent application, employed "two polarizers arranged at right angles to each other and in separate lanterns, in combination with a pair of analyzers [modified opera glasses] having polarizers therein arranged at right angles to each other, the analyzer for one eye being adapted to obliterate the picture produced by one lantern while permitting the picture produced by the other to be seen." By this process, the application continues, "a separate picture is conveyed to each eye, and as these pictures form a stereoscopic pair the impression is irresistibly conveyed in the mind of one picture only, and that is a solid one."[17] An advantage of Anderson's method was that it permitted the projection of

17 John Anderson, "Method by Which Pictures Projected upon Screens by Magic Lanterns Are Seen in Relief." U.S. Patent No. 542.321 (July 9, 1895): 3–4.

full-color images in stereo, but it required the use of a silver "non-depolarizing" screen to ensure the specific linear axis of the polarized images. (Silvered screens would become common in theaters during the Hollywood boom of 1953–54, which popularized polarized 3D motion pictures; see Chapter 3.)

Two anaglyphic systems appeared in France at approximately the same time. In 1896, Paul Mortier patented a reversible stereoscopic camera called the "Alethoscope," which he described as follows: "Coloured filters in complementary colours must be placed in front of the projection lenses, and the spectators must be provided with similarly coloured spectacles." Mortier also envisioned equipping the glasses with alternating, electrically driven shutters to be synchronized with twin projectors. Claude Louis Grivolas, working with entrepreneur Charles Pathé between 1897 and 1901, developed a system that printed twin-strip stereo photographs onto a single band of film with alternating left- and right-eye views. Using red and blue filters and a four-part shutter in his projector, spectators used red and blue glasses to view the alternating frames. Apparently, neither Mortier's nor Grivolas's systems were exploited commercially.[18]

Little took place in the evolution of three-dimensional cinema in the first decade of the twentieth century, but, according to R. M. Hayes, on June 10, 1915, Famous Players Film Company (soon to combine with the Lasky Feature Play Company and become Paramount Pictures, the first vertically integrated studio in the United States) demonstrated three one-reel shorts in anaglyphic duo-color stereo to a private audience—*Jim the Penman* (scenes from a forthcoming Famous Players feature), *Niagara Falls*, and *Rural America* (scenes of New Jersey farmland). Shot with a dual-camera unit designed by pioneering Edison Company director Edwin S. Porter and William E. Waddle, the films were projected in stereo pairs printed one on top of the other and viewed through lorgnette-style colored glasses that decoded the left- and right-eye image. Adolph Zukor, Paramount's founder and president, later claimed that the camera rig used to make the company's 3D features in the early 1950s was based on the Porter and Waddle 1915 version.[19]

---

18 Zone, *Stereoscopic Cinema*, 70–71.

19 R. M. Hayes, *3-D Movies: A History and Filmography of Stereoscopic Cinema* (Jefferson, NC: McFarland, 1989), 4–5. Apparently, Charlie Chaplin also experimented with 3D, or at least his cinematographer Rollie Totheroh did. There are four reels among the "Unknown Chaplin" collection at the British Film Institute that show attempts to demonstrate a stereoscopic effect achieved through "oscillatory projection." Discovered by British inventor Theodore Brown, this was a glasses-less process that moved the camera backward and forward to create a central 3D perspective, in which the edges of the frame were sometimes out of focus. Similarly, the cinematographers for Lon Chaney's *The Hunchback Notre Dame* (Wallace Worsley, 1923) seem to have experimented with a two-camera stereo process although the footage never made it into the completed film.

The novelty period of anaglyphic 3D films

The year 1922 was a watershed for three-dimensional cinema. The first duo-colored anaglyphic feature appeared in September. Shot with a dual-camera unit designed by Harry K. Fairall and Robert F. Elder, *The Power of Love* was directed by Nat Deverich and presented in both Los Angeles and New York City by dual projection with the two stereo prints either tinted in different colors or shown through differently colored filters. The film may later have been projected in a single-strip 3D format, and it had a brief 2D afterlife. A writer for *Popular Mechanics* at the Los Angeles premiere described *The Power of Love* as "a motion picture in which the characters seemed to be moving about in locations which had a depth exactly like the real spots where the pictures were taken," but the film has since been lost.

In December 1922, Educational Films released *Plastigrams*, one of several 3D shorts made over the next year by the team of Jacob F. Leventhal and Frederick Eugene Ives in a dual 35mm single-strip anaglyphic process. Playing continuously in New York City for nearly two years, *Plastigrams* was shot by D. W. Griffith's cameraman Billy Bitzer and rereleased in September 1924 with a DeForest Phonofilm optical soundtrack, making it the first three-dimensional sound film. (In 1925, Leventhal and Ives were licensed by Pathé to distribute a series of six more anaglyphic shorts with Phonofilm accompaniment under the collective title of *Stereoscopiks*; photographed by William T. Crespinel, these were shown exclusively at New York's Rivoli Theater and were notable for their optical effects of things emerging into audience space directly from the screen.) Simultaneously, William Van Doren Kelley, branding the same system as "Plasticon" but using different equipment, produced the two-reel show *Motion Pictures of the Future*, consisting of *Plasticon* (a demonstration of the 3D process itself) and *New York City* (a travelogue that exploited views from the tops of skyscrapers). Among other wonders, *Plasticon* showed the unique possibility of having two different endings to the same film by using different color coding for each scene (Figure 2.1).

Also in December 1922, inventor Laurence Hammond introduced "Teleview" at the Selwyn Theater in New York City. This was an eclipse system, designed by William F. Cassidy, using a twin-strip camera with two lenses and twin projectors synchronized to alternate left- and right-eye frames; the images were viewed through a special device containing a revolving electrical shutter, which was attached to the armrest of each patron's chair. The 85-minute program included short documentaries on the Hopi Indians and the Canadian Rockies, as well as the science fiction feature *M.A.R.S., or Radio Mania*, shot by veteran Hollywood cinematographer George Folsey. The film critic for the *New York Times* thought *M.A.R.S.* was tedious as drama, but wrote of the Teleview process

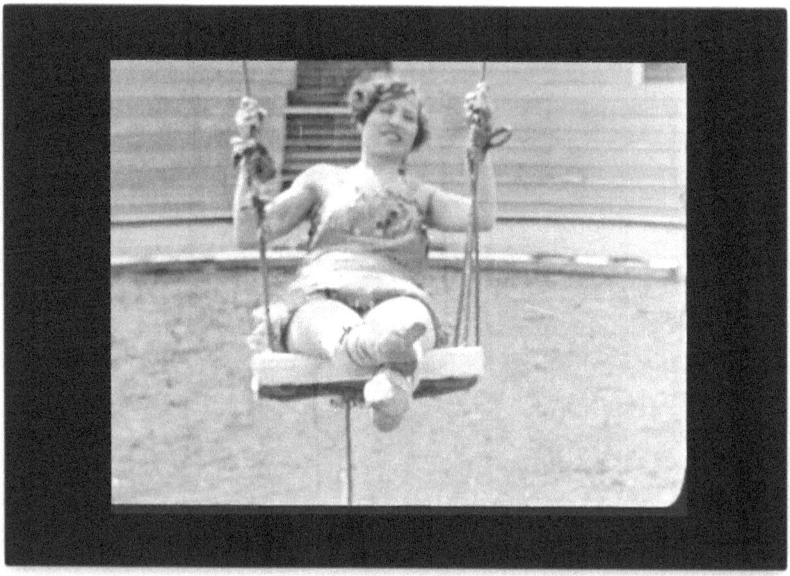

Figure 2.1 From the "Plasticon" demonstration in *Motion Pictures of the Future* (William Van Doren Kelley, 1922): the first widely seen stereoscopic film, shot in dual-lens 35mm single-strip anaglyphic 3D and two-color Technicolor.

itself: "Stereoscopic motion pictures have come [...] The impression of objects in relief that one gets from them is startling, so real that it seems inappropriate to speak of it as an illusion." Despite this enthusiasm, Teleview was too impractical for mass diffusion, and it disappeared after its initial run at the Selwyn.[20]

That 3D cinema was distinctly in the air by the mid-1920s is demonstrated by the fact that Abel Gance (1889–1981) had anaglyphic footage shot for his epic *Napoléon vu par Abel Gance* (1927), which was intended to be, among other things, a compendium of all existing film techniques. Most striking among these was a widescreen process that anticipated Cinerama by twenty-five years in that it employed a triptych, or three-panel screen, to show three standard 35mm images side by side. This image was photographed by three identical Parvo-Debrie cameras mounted one on top of the other in an arc and synchronized to run concurrently by means of a flexible motor shaft. (Not only did Polyvision foreshadow Cinerama, but in 1941 it gave Professor

20 Zone, *Stereoscopic Cinema*, 107. In *Foundations of Stereoscopic Cinema: A Study in Depth* (New York: Van Nostrand Reinhold, 1982), Lenny Lipton notes that fifteen 3D films were released during the 1920s. Some of these were scientific and/or marketing shorts, but all except those noted above are lost.

Henri Jacques Chrétien the idea for inventing the anamorphic lens, the basis of the CinemaScope and Panavision systems in the 1950s.) Gance used other cameras for color and for stereoscopic photography—apparently, dual 35mm printed single-strip anaglyphic—several reels of which were shot by his cinematographer Leonce-Henry Burel, but neither the color nor the 3D footage made it into the final cut because the director felt they detracted from the narrative. As Gance later recalled:

> To see the rushes, I had to wear those red and green spectacles. The 3D effects were very good, and very pronounced, I remember one scene where soldiers were waving their pistols in the air with excitement, and the pistols seemed to come right out into the audience. I felt, however, that if the audience saw this effect, they would be seduced by it, and they would be less interested in the content of the film. And I didn't want that at all.[21]

British filmmaker and historian Kevin Brownlow, who has devoted much of his career to assembling various restorations of *Napoléon* (the most recent in 2017), has said of the film: "The visual resources of the cinema have never been stretched further [...] The picture is an encyclopedia of cinematic effects—a pyrotechnical display of what the silent cinema was capable of in the hands of a genius."[22] And yet, Gance ultimately felt that the stereoscopic effect should not be among these, suggesting that its novelty status as spectacle prohibited its use in a serious narrative film (even an epically spectacular one).[23] This is the paradox of three-dimensional cinema—its immersive impact has historically relegated it to genre cinema, like action-adventure and horror. Even today, every serious venture in the format is outmatched at the box office many times over by adaptations from the Marvel Cinematic Universe and other comic serials, which are unthinkable to their producers—perhaps because potentially unmarketable—without the embellishment of digital 3D, whether native or converted in post-production.

In the United States, the novelty period of three-dimensional cinema extended into the 1930s and 1940s, often combined with the somewhat newer technologies of natural color and optical sound. In January 1936, for example, MGM released the seven-minute *Audioscopiks* as a short subject preceding

21 Quoted in Kevin Brownlow, *The Parade's Gone By* (New York: Bonanza Books, 1968), 559.

22 Quoted in David A. Cook, *A History of Narrative Film*, fourth edition (New York: W. W. Norton, 2004), 308.

23 Ironically, the 3D sequence was to have been placed just before the Intermission, and it seems Gance was concerned that its immersive impact would have detracted from the rest of the spectacle, especially the Polyscope triptychs of the second half.

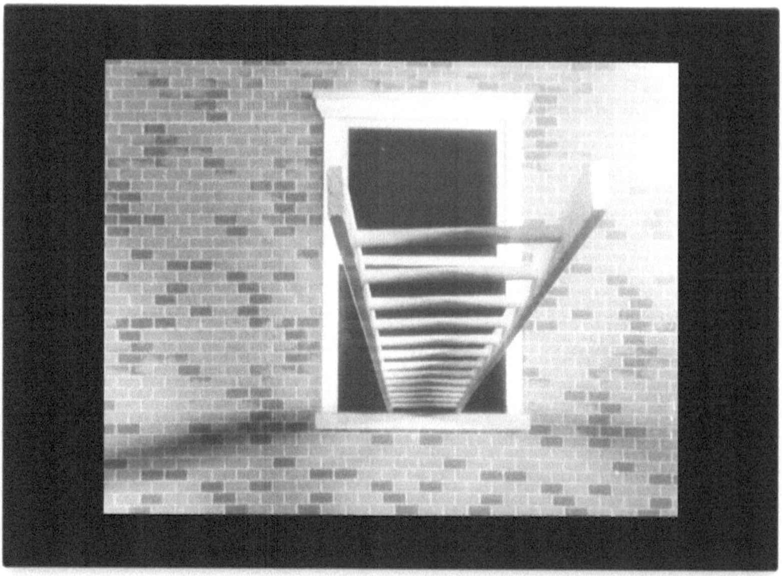

Figure 2.2 From *Audiosopiks* (John A. Norling/Jacob F. Leventhal, 1936): dual-lens 35mm single-strip anaglyphic 3D, two-color Technicolor, and Western Electric Mirrophonic Sound.

its acclaimed black-and-white 2D feature *A Tale of Two Cities* (Jack Conway, 1935). Shot with a dual-lens, single-strip anaglyphic camera designed by John A. Norling and *Plastigrams* director Jacob F. Leventhal for viewing through lorgnette-style red and green spectacles, the film featured two-color Technicolor and Western Electric Mirrophonic sound,[24] and its display of 3D imagery included a Spanish flamenco singer, a slide trombone, a ladder emerging from a window, a girl on a swing, and a baseball thrown directly at audience, as well as a demonstration of the Norling-Leventhal stereoscopic camera system (As in the eras of early cinema and early widescreen, the operation of the apparatus was foregrounded as an attraction during the novelty period of 3D m otion pictures.) (Figure 2.2). The short was so successful that it generated a sequel, the eight-minute *New Audioscopiks* (1938), another compilation of footage shot with the Norling-Leventhal camera. A third MGM short, *Third Dimensional Murder* (1941), was shot by George Sidney, who would later direct the studio's

24 "Mirrophonic" was the tradename for Western Electric's variable density optical sound system from the late 1920s through the 1950s. It was a nonlinear system similar to Fox Movietone and had higher fidelity than variable area linear systems, such as RCA Photophone, at the time.

popular 3D musical *Kiss Me Kate* (1953), and photographed with a new two-camera rig built by John Nicklaus and branded as "Metroscopix." The film was a parody of Universal's popular *Frankenstein* series in which the narrator is called to a remote gothic castle and attacked successively by a skeleton, a murderous Indian warrior, and the Frankenstein monster. In 1953, MGM bundled all three shorts into a 24-minute package for distribution under the title *Metroscopix* to capitalize on the 3D boom of that year.

In their years of distribution as individual short subjects, *Audioscopiks*, *New Audioscopiks*, and *Three Dimensional Murder* became the most widely seen 3D films of all time, but there were complaints that the anaglyphic process did not create a sensation of real depth, offering rather, as one critic put it, "a series of thin, cardboard models placed in different planes" that lacked "roundness." There was also the nagging complaint of eye strain caused by the color-coded glasses. Neither of these issues had arisen with stereoscopic still photography because (1) the images did not have the vector of motion to contend with and (2) the individual viewing time per image was of short duration. Camera engineer John A. Norling, in an article that appeared in the *Society of Motion Picture Engineers Journal* in 1939, defended anaglyphic stereoscopy: "It has been claimed that the anaglyph system, because one eye sees one color and the other its complementary, causes serious retinal rivalry [...] but it has been my observation that no one suffers any discomfort in viewing a reel or two."[25] At the same time, he acknowledged that polarization—harking back to John Anderson's turn-of-the-century research and acknowledging the contemporary work of Edwin H. Land—was a superior method for stereoscopic image selection because it allowed for full-color projection and provided a "more complete elimination of the unwanted image."[26] Thus, even while promoting his proprietary anaglyphic system, Norling accurately forecast that the future of three-dimensional cinema lay in the polarized image.

The advent of polarized 3D films

In 1932, while he was still an undergraduate at Harvard, Land developed a cost-effective light polarizing material that he hoped to sell to the automobile industry as a means of reducing headlight glare, but he was also interested in its application for stereoscopic cinematography since polarizing filters could serve the same function as anaglyphic ones, with the added advantage of providing full-color imagery. Fundamentally, a polarized filter acts like a comb that

---

25 J. A. Norling, "Three Dimensional Motion Pictures," *Society of Motion Picture Engineers Journal* (December 1939): 612.

26 Quoted in Zone, *Stereoscopic Cinema*, 149.

permits light to pass through only if its waves are oriented in the direction of the lines of the filter. (Land himself used the analogy of a picket fence through which sticks can pass only if they are on the same axis as the pickets.) In January 1936, Land formed the Polaroid Corporation with George W. Wheelwright III and Clarence Kennedy, and they later gave a successful public demonstration of their process at the Waldorf Astoria Hotel in New York. In December 1936, a polarized 3D movie produced by Land opened at the New York Museum of Science and Industry and played to large crowds for several years after. In Italy that same year, Societa Italiana Stereocinematographia released the first polarized 3D feature, *Nozze Vagabonde* (*Beggar's Wedding* [Guido Brignone, 1936]) in a black-and-white single-strip process, and the following year in Germany Ziess-Ikon presented the Raumfilm-System commercial short *Zum greifen nah* (*You Can Almost Touch It* [Kurt Engle, 1937]), the first polarized 3D film shot in color (on newly developed Agfacolor stock).

In the United States, John A. Norling and Polaroid were commissioned by the Chrysler Corporation to make a 15-minute 3D short as the main attraction of its exhibit at the 1939 World's Fair. *In Tune with Tomorrow* was initially shot in black-and-white with two 35mm Bell and Howell cameras mounted so that one was upside down. (The point of this was for the two lenses to be brought close enough together to approximate the interaxial separation of 2½ inches between the human eyes.) In its first year, the film was viewed by 1.5 million attendees, and it was the most elaborate stereoscopic film made to date, whose content was described by Norling as the "actual operation of the half-mile long Plymouth Plant in Detroit and a trick-assembly picture that shows the Plymouth car coming into being without the apparent aid of human hands."[27] The automobile assembly sequence used a stop-motion technique in which the thousands of separate manufacturing parts were animated one frame at a time and took 36 days to shoot. With generous funding from Chrysler, Norling's film was remade as *New Dimensions* in three-color Technicolor the following year and remained one of the Fair's most popular and spectacular attractions. Seen by another 4.5 million people at the Chrysler exhibit, it was picked up for nationwide theatrical distribution by RKO Pictures in 1953 (Figures 2.3 and 2.4). (Norling subsequently made the promotional documentary *Thrills for You* for the Pennsylvania Railroad exhibit at the 1940 Golden Gate International Exposition in San Francisco, a black-and-white 3D short that included footage taken from the front of a rapidly moving locomotive and dynamic shots of on-rushing trains.)[28]

---

27 Ibid., 156.

28 At the same 1939 World's Fair, Hollywood special effects artist Fred Waller (1886–1954) debuted an immersive theater for the Petroleum Industry exhibit called Vitarama

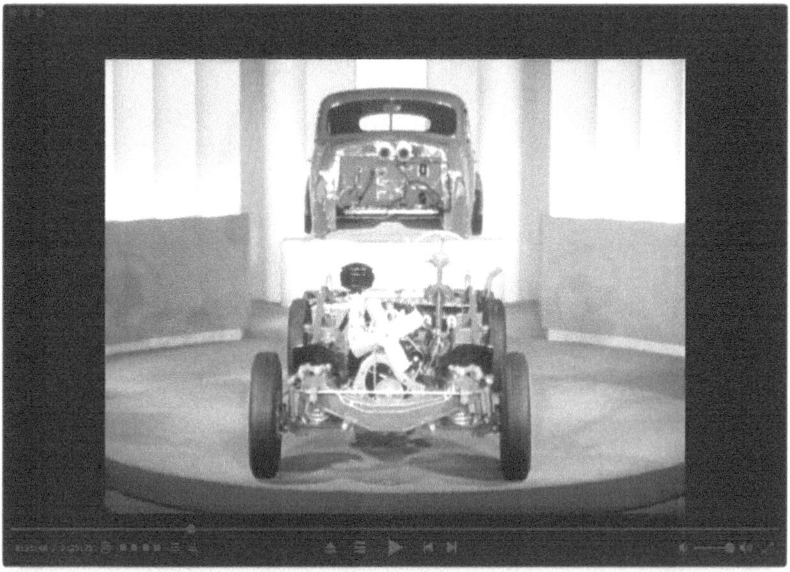

Figure 2.3 From the stop-motion Plymouth Sedan assembly in *New Dimensions* (John A. Norling for Chrysler Corporation Exhibit, 1939–1940 World's Fair): dual-camera polarized 3D; three-color Technicolor, optical sound.

During the 1940s, a 3D system was developed in the Soviet Union called Stereofilm that employed a so-called "lenticular" screen for separating polarized stereo pairs without the aid of glasses. This screen used a projection surface composed of rasters—in this case, thousands of flattened copper wires, regularly spaced in parallel and angled to reflect the stereo pairs back to the audience to create a three-dimensional image. Invented by Russian filmmaker Semyon Ivanov, Stereofilm is technically characterized as an "autostereoscopic parallax stereogram" system that uses the screen as a kind of "perspective

that used an array of 11 projectors to create a giant image on a dome-like screen. Two years later, he invented the Waller Flexible Gunnery Trainer for the military, a multiprojection simulator to train airplane gunners that used the same technology as Vitarama, together with audio and vibration feedback; its first installation was in Honolulu, just after the Pearl Harbor attack. In the early 1950s, Waller used the same idea to create multicamera, multiprojector Cinerama, the catalyst for the industry-wide conversion to widescreen. This system used three interlocked 35mm cameras mounted together at 48-degree angles to shoot a 146-degree-wide image, similar to Abel Gance's Polyvision triptychs (see above), which was then spread across a large curved screen by three synchronized projectors to the accompaniment of seven-track stereophonic surround sound.

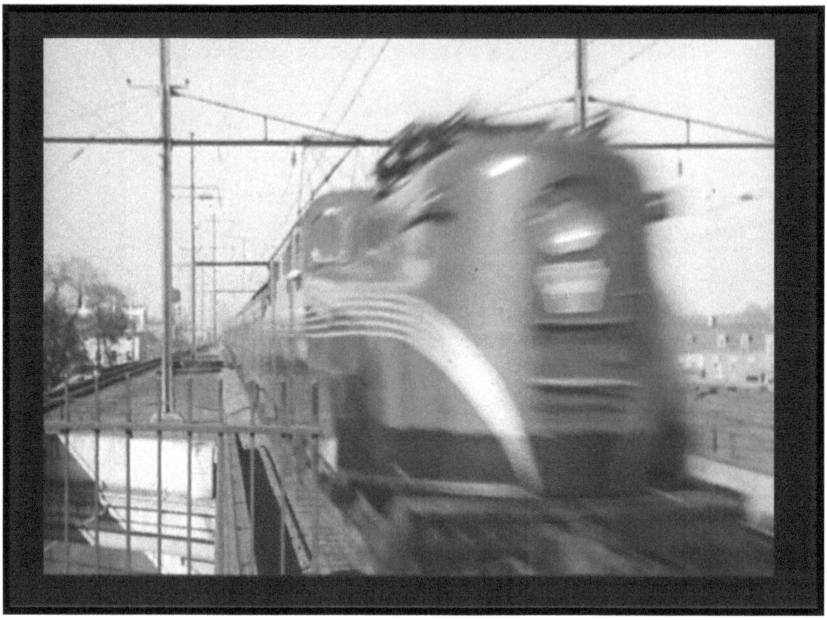

Figure 2.4 From *Thrills for You* (John A. Norling for the Pennsylvania Railroad Exhibit, 1940): dual-camera polarized 3D, black-and-white, optical sound.

grille."[29] Its effect was impressive but limited to a narrow cone of vision so that any head movement by the spectator would destroy the illusion. The first Stereofilm production was *Kontsert* (*Concerto*; aka *Land of Youth* [Aleskandr Andreyevskiy]), shot in Sovcolor, which opened at the specially constructed 180-seat Moscow Stereokino theater in February 1941. This venue presented autostereoscopic movies on a grid screen for another 18 years and was later joined by four others throughout the country. The Second World War briefly stalled the Soviet film industry due to the Nazi invasion, but a number of Stereofilm features were produced immediately thereafter, including *Robinson Kruzo* (*Robinson Crusoe* [Aleskandr Andreyevskiy, 1947]), photographed in Sovcolor on 70mm film. On seeing it, the great Russian film director and theorist Sergei Eisenstein (1898–1948) wrote shortly before his death, "To doubt that stereoscopic cinema has its tomorrow is as naïve as doubting whether there will be a tomorrow at all."[30]

The war postponed the commercial exploitation of three-dimensional cinema in much of the West as well, but in 1951 an exhibition in London

29 Zone, *Stereoscopic Cinema*, 168.

30 S. M. Eisenstein, "About Stereo Cinema," trans. Catherine de la Roche, *Penguin Film Review* 8 (January 1949): 35.

pointed in that direction. The five-month-long Festival of Britain opened on the south bank of the Thames in May of that year, marking the 100th anniversary of the Crystal Palace exhibition that had been a showcase for Victorian invention—including, of course, stereoscopic photography. Appropriately, stereoscopic motion pictures became a main attraction a century later, as the Festival featured a futuristic 400-seat theater called the Telecinema that provided dual 3D projection in full color with interlock stereophonic sound, as well as live video projection. Sponsored by the British Film Institute and under the technical direction of Raymond Spottiswoode, the Telecinema offered a program of four stereoscopic shorts, two of which were produced by the experimental animator Norman McLaren of the National Film Board of Canada (NFB): *Now Is the Time (To Put on Your Glasses)* exploited McLaren's famous technique of painting/drawing directly onto the film and featured a novel Animated Sound system in which he photographed waveforms directly onto the soundtrack to create a synthetic score; *Around and Around* relied on the filming of moving oscilloscope patterns with a similarly photographed soundtrack, both films creating electronic music that predated the Moog synthesizer by several decades.[31] The other two 3D films were *Royal River* (initially titled *The Distant Thames*), a documentary photographed in long shots with two three-strip Technicolor cameras from a floating platform, and *A Solid Explanation*, a black-and-white demonstration of the principles of stereoscopy intercut with scenes of various animals in the London Zoo. Based on their experience with the Telecinema, Raymond Spottiswoode and his brother Nigel subsequently formed a company called Stereo Techniques, Ltd., and they published a comprehensive book on stereoscopic filmmaking called *The Theory of Stereoscopic Transmission and Its Application to the Motion Picture* (Berkeley: University of California Press, 1953). Stereo Techniques itself went on to produce a series of 18 shorts for theatrical distribution in the United Kingdom using a 3D rig called Spacemaster, as well as the only British 3D feature of the 1950s *The Diamond Wizard* (Mongomery Tully and Dennis O'Keefe, 1954); some of these shorts and the feature were later released in the United States. According to R. M. Hayes, it was the theoretical writings of Raymond Spottiswoode—and not the Hollywood trade paper *Variety*—that coined the term "3-D" to refer to three-dimensional cinema.[32]

31 The first feature-length film to use an entirely electronic score was MGM's *Forbidden Planet* (Fred M. Wilcox, 1956); it was composed by Bebe and Louis Barron, who were credited with "Electronic Tonalities" because they were not members of the Musician's Union.

32 Hayes, *3-D Movies*, 19.

Chapter 3

"SEE IT IN 3 DIMENSION!": THE FIRST HOLLYWOOD 3D BOOM, 1952–55

Hollywood at the brink

The year 1948 was a terrible one for the American film industry: television was spreading through the country like a plague, the anti-Communist witch hunt and blacklist reared their ugly heads, and the Supreme Court had ordered the demonopolization of the studio system in the *United States* v. *Paramount* decision, resulting in the "Paramount Decrees" in May.[33] In 1946, weekly motion picture attendance peaked at an all-time high of 100 million, or two-thirds of the population; by the end of 1948 it had declined to 75 million—the lowest number since the Great Depression—and in 1952 it declined still further to half the 1946 figure. It was all bad news, but the worst of it was the little black-and-white box called television. When it was introduced as a mass entertainment medium in the United States just before the Second World War, there had been argument among its promoters as to whether it should be marketed as "radio with pictures" or "movies in the home." After a brief hiatus to accommodate wartime manufacturing restrictions, they decided to go with the latter—going in head-to-head competition with the world's most powerful leisure time industry, and by 1948 they were starting to win: in that year 1 million Americans owned television sets, and two years later there were ten times that many; general diffusion of the product followed during the next decade, and by 1960 market saturation was complete. After some fumbling attempts to buy up the TV networks (foiled by the Supreme Court's 1948 divestiture order), the Hollywood majors saw that they would have to fight the small-box cancer on its own terms—with image size, sound, and color.

---

33 These were consent decrees that forced the five major studios (MGM, Warner Bros., Paramount, 20th-Century Fox, and RKO) to divest themselves of their lucrative theater chains according to a mutually agreed-upon schedule over the next five years. The Supreme Court held that, together with the minor studios (Universal, Columbia, and United Artists), they had conspired to monopolize motion picture production, distribution, and exhibition in the United States from 1934 to 1937.

First, since television was still a black-and-white medium, the film industry attempted a conversion to full-color production, stimulated by a conversion from three-color Technicolor to far cheaper single-strip Kodak Eastmancolor stock. (The now notorious fading problems with Eastmancolor nearly caused a generation of 1950s color films to be lost until digital restoration became possible in the 1990s.) In 1947, only 12 percent of American features were made in color; by 1954 the figure had risen to more than 50 percent, and by 1975 to 95 percent. Next, in a simultaneous attempt to exploit the *size* of the cinema screen, Hollywood began to experiment with new optical systems that lent greater width and depth to the image.

Cinerama

The earliest of the new formats was a multiple-camera/projector system branded as Cinerama that revived the Polyvision process Abel Gance had used in *Napoléon* (notably, at the expense of 3D; see Chapter 2) in which three synchronized 35mm cameras linked together in an arc would simultaneously record a wide-field image. This would later be cast by three similarly linked projectors onto a vast wraparound screen, creating what appeared to be a single image but was in fact a three-screen triptych. The seams between the film strips were concealed by a slight overlapping of the camera lenses and floating metal combs in the projectors, a solution that was never wholly satisfactory. Still, the Cinerama image was six times the standard size, and its deeply curved 146-degree screen enhanced peripheral vision; it also surrounded the audience with seven-track (initially six-track) stereophonic sound, recorded magnetically rather than optically, providing an immersive sonic experience appropriate to its enormous screen. Debuting with *This Is Cinerama* in 1952, the process was cumbersome and expensive for both producers and exhibitors (converting a theater to Cinerama cost about $75,000), and despite its initial popularity, it never really proved viable for the production of theatrical features. Like IMAX in its analog phase, Cinerama worked best for spectacular documentaries that took audiences on wild roller-coaster rides (of course), overflights of dynamic scenery, and exotic sea-faring adventures. Only a handful of dramatic features were made in the multiple-camera/projector process (e.g., *How the West Was Won* and *The Wonderful World of the Brothers Grimm*, both 1962), and in 1963 Cinerama converted to a single-lens wide-film widescreen system based on Todd-AO (see below). Cinerama is important not because it lasted but because it brought audiences back into the theaters in large numbers for the first time since 1946, proving that an immersive screen experience could successfully compete with the little black-and-white image boxes that had invaded America's homes.

Natural vision

In Hollywood, even as the Festival of Britain was taking place, two cameramen named Lothrop Worth and Friend Baker had begun to experiment with 3D, building a 16mm prototype for a dual-camera, lens-to-lens 3D system. Screenwriter Milton Gunzberg, impressed with their test footage, convinced his brother Julian, an ophthalmologist, to form a corporation to develop a 35mm dual-camera rig to produce 3D documentary shorts. Worth and Baker, working with cinematographer Joseph Biroc and O. S. Bud Byrne, built the rig—which involved two cameras facing an angled mirror—and the Gunzbergs patented their process as Natural Vision 3-Dimension, emphasizing that its convergence feature replicated human binocular vision. In early 1952, the Natural Vision Corporation began screening test footage for Hollywood studio executives, but none were impressed. However, radio and independent film producer Arch Oboler was making a low-budget African adventure feature called *The Lions of Gulu*, and he decided to invest $10,000 of his own money to partner with the Gunzbergs and shoot the film in Natural Vision and Anscocolor.[34]

Retitled *Bwana Devil*, it was filmed on location in the Malibu Hills[35] with Worth as the camera operator and Biroc as director of photography. Released on November 26, 1952, *Bwana Devil* was an immediate hit, grossing nearly $100,000 in its first week in two theaters, and within days Jack Warner had licensed the Natural Vision process from the Gunzbergs and commissioned Lothrop Worth to shoot *House of Wax* (Andre de Toth, 1953), the first mainstream Hollywood feature to be photographed and released in 3D.[36] (Lathrop ultimately became camera operator, with the DP credit going to

34 A single-strip three-color process based on German Agfacolor (and in fact merged with that company as Agfa-Anscocolor from 1928 to 1939, when it became General Analine & Film (GAF). After the Second World War, the corporation operated in the United States as Anscocolor but did not completely shed its German roots until 1967.

35 *Bwana Devil* also contained 2D inserts of documentary footage shot in Africa several years before.

36 R. M. Hayes, *3-D Movies: A History and Filmography of Stereoscopic* Cinema (Jefferson, NC: McFarland, 1989), 20–21. Technically, "The first 3-D feature produced by a major studio," as its ad copy read, was Columbia's low-budget film noir *Man in the Dark* (Lew Landers, 1953). Begun as a 2D remake of Columbia's 1936 *The Man Who Lived Twice* (Harry Lachman), the picture was converted to 3D when studio engineers assembled a twin Mitchell camera rig to produce stereo pairs: Columbia boss Harry Cohn had the film reshot in 11 days and rushed it into release two days before the opening of *House of Wax* to scant commercial success and little critical notice. In fact, *Man in the Dark* had some genuinely original emergence effects that simulated the hallucinatory disorientation of the film's hero after brain surgery, as can be seen in Columbia's 2014 Blu-ray restoration.

Hollywood veteran Burt Glennon, assisted by future Hitchcock collaborator Robert Burks.) (Figure 3.1).

A remake of *The Mystery of the Wax Museum* (Michael Curtiz, 1933), originally shot in the novel two-strip Technicolor process, *House of Wax* featured not only full-fledged Technicolor 3D imagery but also stereophonic sound to provide an illusion of acoustic depth to match the visuals. WarnerPhonic Stereo was a four-channel system that used a 35mm magnetic track for the left, center, and right speakers behind the screen and a monaural optical track for the effects channel. When it premiered at the 3,600-seat New York Paramount Theater on Times Square on April 10, 1953, a total of 25 speakers had been installed throughout the auditorium to produce fully directional sound and surround effects that immersed the viewer in the action. Spectacle notwithstanding, *House of Wax* was an intelligent, entertaining, and well-mounted horror film, starring Vincent Price as an ingeniously murderous sculptor, and it took in an amazing $531,000 at the Paramount alone, nearly amortizing its negative cost of $680,000 in 40 days. It moved *Variety* to write in December that 3D was the film industry's "Next Big Thing," and columnists were reporting that every studio in Hollywood

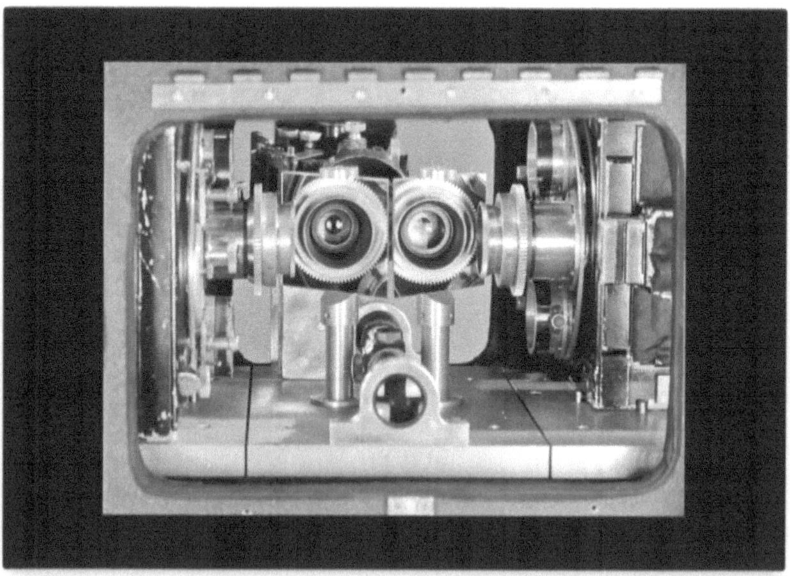

Figure 3.1 Dual Camera/Lens Configuration of a Natural Vision Camera, *c.*1952 (Natural Vision Corporation, 1953).

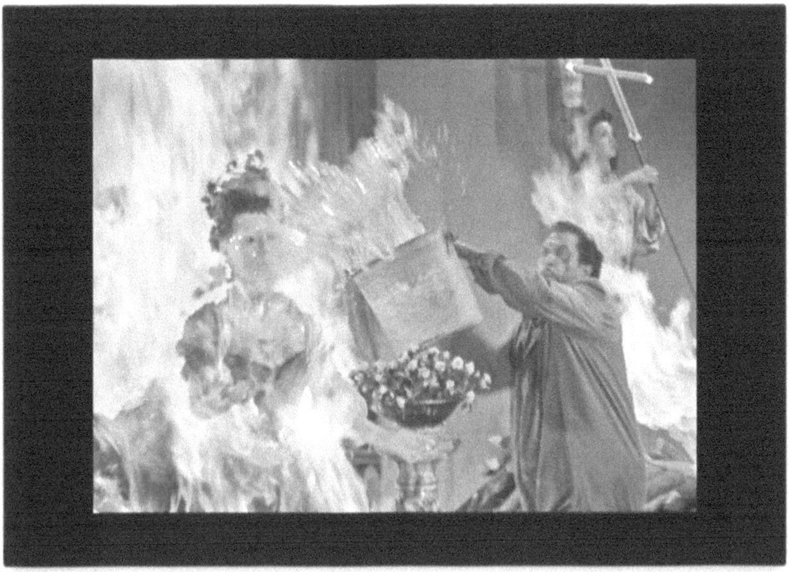

Figure 3.2 *House of Wax* (Warner Bros., 1953): the first Hollywood feature to be shot in Natural Vision 3D, with three-color Technicolor (hereafter, designated as "Technicolor") and WarnerPhonic Stereo sound. Vincent Price as Dr. Charles Jarrod attempting to extinguish the fire in his original wax museum. Shown here in its original aspect ratio of 1.37:1.

was working on some sort of 3D system to produce what the trade press now called "depthies."[37] Warner's and Columbia had both signed contracts with Natural Vision, while Paramount, MGM, and Universal declared that they would make 3D films using their own proprietary rigs. Warner's, in fact, prematurely announced that *all* of its films would henceforth be produced in Natural Vision, and over 500 theaters nationwide had contracted for the installation of WarnerPhonic sound. (In fact, only five of the studio's next 20 films were released in 3D.) (Figures 3.2 and 3.3).

---

37 Hal Morgan and Dan Stymmes, *Amazing 3-D* (Boston, MA: Little, Brown, 1982), 56, 65. This heavily illustrated book also contains a detailed chapter (4. "Superman Breaks Loose, pp. 106–54) on the production of anaglyphic 3D comic books during the early 1950s. These appeared in a wide range of genres, including war, crime, horror, romance, and superhero, and they sold over 2 million copies between 1953 and 1954, when their popularity waned.

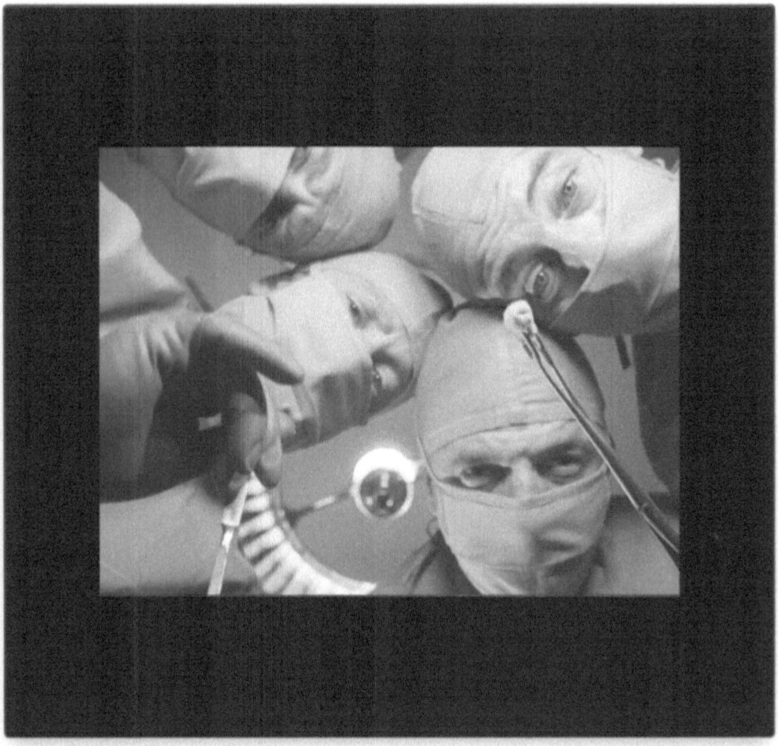

Figure 3.3 *Man in the Dark* (Columbia Pictures, 1953): released two days before *House of Wax* so it could claim priority, the film was shot in eleven days in black-and-white with a twin Mitchell camera rig to produce its stereo pairs. Here, stereoscopic brain surgery from the patient's point of view.

CinemaScope

At the same time, 20th Century Fox shot its first film in a new, non-stereoscopic format called CinemaScope, which was advertised as "The Modern Miracle You See without Glasses," stressing the new industry emphasis on depth. This system was based on the "Hypergonar" anamorphic distorting lens invented by Henri Jacques Chrétien in 1928 that squeezed a wide-field image onto conventional 35mm film stock and restored it as a widescreen image with a compensating lens in projection. The aspect ratio of the film frame to this point, multiple-camera/projector systems notwithstanding, had been standardized in 1932 by the American Academy of Motion Picture Arts and Sciences at 4:3 or 1.33:1 (in reality, 1.37:1, to include room for the optical soundtrack). CinemaScope offered a radically new ratio of 2.55:1 (about 8:3, later adjusted downward to 2.35:1), slightly less than that of Cinerama, and it similarly gave the image a broadly oblong shape and enhanced peripheral vision when projected on a curved screen. The process, introduced to the public in September 1953 in *The Robe* (Henry Koster), featured a four-track stereophonic soundtrack magnetically recorded on the film strip, and it was aggressively marketed as a cost-effective alternative to both 3D and Cinerama, with the natural advantage that it was a single-camera/projector system, requiring only special lenses, a metallized wide screen, and a four-track stereophonic sound system, rather than multiple cameras and projectors. (Fox advertised a conversion cost of $10,000 to $12,000 for smaller theaters, plus $2,800 for a pair of anamorphic lenses, which would drop to $1,000 a pair by April 1954.)[38] For ideal exhibition, 3D also required a metallic screen, but a more expensive one than CinemaScope: polarized filters lowered light levels (think of the example of sunglasses), so that 3D screens had to be made more reflective by beading them with silver particles. Such screens were highly directional and caused sideline seating to become unusable for both 3D and regular films due to their angular darkening. Later 3D films like *Kiss Me Kate* (George Sidney, 1953) that opened in wide-seat, widescreen venues such as Radio City Music Hall were often projected flat to increase audience capacity.

For nearly a year, 1953–54, CinemaScope and 3D competed in Hollywood as rival optical systems, and both had been demonstrated to 4,600 of the nation's exhibitors on the same weekend in March 1953. Although Natural Vision and its imitators were not widescreen processes as such, films produced in 3D were often cropped for projection in aspect ratios ranging from 1.50:1 to 1:1.85. This was done to accommodate the wider screens that exhibitors were installing in their theaters for true widescreen formats like CinemaScope and VistaVision (see below) and to cater to the growing public taste for a wider image. Thus both anamorphic and stereoscopic movies, somewhat speciously in both cases, could claim to offer audiences the same optical properties of width and depth. According to Bob Furmanek, the "widescreen, three-dimensional and stereophonic sound revolution […] spread like wildfire in the spring and summer of 1953," so that 1.37:1 "was rapidly declining as a compositional ratio among filmmakers" by June of that year.[39] Even so, studios announced that new product would be "suitable for all screens," because exhibitors who couldn't convert their theaters quickly enough feared a product shortage and those who had already done so knew that standard ratio features

38 Ray Zone, *3D Revolution: The History of Modern Stereoscopic Cinema* (Lexington: University Press of Kentucky, 2012), 72.

39 "The First Years of Widescreen Production," http://www.3dfilmarchive.com.

would look wrong on their wider screens. This meant that not only would 3D films be cropped during the widescreen conversion but so too would the backlog of nearly 200 unreleased studio features already shot in the 1.37:1 ratio, producing some compositional disasters—most notoriously George Stevens's *Shane*, which Paramount cropped to 1.66:1 and projected through distortion-producing wide-angle lenses for its April 23, 1953, premiere at Radio City Music Hall.

Parallax, convergence, and focal length

As stated earlier, depth perception depends on parallax, the displacement in the apparent position of an object viewed along the slightly different lines of sight provided by our eyes. The stereoscopic effect in movies consists of two distinct parallax phenomena: ***positive parallax***, in which objects/characters appear to recede in a three-dimensional perspective from the surface of the screen (the "peepshow effect"), and ***negative parallax***, in which objects/characters appear to loom into the space of the auditorium from the surface of the screen (the "emergence effect"). In both cases, the point of demarcation for the parallax is the flat, empirically two-dimensional screen.

The key to parallax in 3D filmmaking is setting the distance between the left and right cameras, a lens separation known as *interaxial distance*. The interaxial, or IA, may be the same as that between the human eyes, or about 65mm (2.5 inches), or sometimes greater. The most sophisticated 3D camera rigs of the 1950s could alter the interaxial to less than 65mm, which helped in managing the total amount of screen parallax, but this depended on the design limitations of any given system. The ability to shift the interaxial from one shot to the next was an important technical achievement that could become a powerful artistic tool. Natural Vision, which was used to shoot ten 1950s features, had a fixed interaxial of 3.5 inches, one inch greater than the probable distance between the eyes, or 2.5 inches, which was exactly that of Charles Wheatone's various stereoscopic experiments of the 1830s. To circumvent undue parallax, Natural Vision films tended to rely on medium shots and long shots, using close-ups sparingly. When such close-ups were essential to the narrative, efforts were made to shoot them in close proximity to a backdrop, for example, a few feet in front of a stockade wall or thick velvet drapes, to avoid severe parallax separation in the background. *House of Wax*, for example, makes frequent use of this technique. In some Natural Vision films, close-ups were filmed against a clear blue sky which, in the absence of clouds, provided no divergent parallax points to cause problems. (Warner's 1953 *The Charge at Feather River* uses this composition a lot.)

The MGM rig, Metro Tri-D, also employed a fixed interaxial, although smaller than that of Natural Vision, so that it could achieve beautiful close-ups that still faced the challenge of divergent background parallax. In some close-ups in MGM's *Kiss Me Kate* (1953), for example, the background is darkened or kept in soft focus to compel audience attention to the foreground, where parallax values are very low and stereo fusion is most comfortable. The Fox rig was an L-shaped configuration employing two Cinesimplex cameras at right angles, plus a single mirror in front of the left and right lens. Some Fox films like *Inferno* (1953) and *Gorilla at Large* (1954) seem to have been shot using a fixed interaxial; and several medium close-ups of *Inferno* display positive parallax values as high as 9 percent of screen width, which is visually challenging for most spectators. The All-Media Camera used for Warner's later 3D pictures had a beam-splitter that could theoretically provide very small lens separations. The studio's *Dial M for Murder* (1954), for example, seems to have combined a slightly reduced but unchanging interaxial for close-quarter filming on a relatively small set to yield meaningful 3D without severe parallax distortion. Paramount's Paravision system used two synchronized Mitchell NC studio cameras mounted on a common base and facing each other; in between was a mirror assembly comprised of two mirrors, each at a 45-degree angle to its respective camera lens. Parallax values in that studio's *Jivaro* (1953) were very nearly perfect, with a maximum possible parallax of 4 percent of frame width—enough to present a deep, solid image but not enough to tax audience vision.

All cameras have an aperture for the F-stop, and the more open it is, the less the focal length of the lens: backgrounds and middle grounds shot with wider apertures lose focus. A smaller F-stop hems the light in, squeezing and pinpointing it, so that the foreground, middle ground, and background of a shot, under the right conditions, can all be in sharp focus. Close-ups are usually shot with an open F-stop, while long shots use a very small one to create a wider field of focus. Focal length has a psychological effect on the way three-dimensional cinema works. It is very easy, for example, for full depth of field in 3D to cause audiences to become distracted by background action. On the other hand, the use of mirrors in most 3D camera rigs of 1950s kept DPs from using very wide angles, and 30mm was the standard lens aperture for most productions. Yet even in the early months of the 1950s boom, professional camera crews were more than capable of making beautiful, sensuous three-dimensional films.

CinemaScope vs. 3D during the year of conversion

By the end of 1953, every major studio in Hollywood except Paramount had been licensed to make anamorphic films—all versions of CinemaScope with proprietary branding (Warnerscope, Metroscope, etc.)—but 3D productions were still in the pipeline. The majors were effectively hedging their bets to see which format would find the most public favor. For example, Warner Bros. announced that *A Star Is Born* (George Cukor, 1954) and *East of Eden* (Elia Kazan, 1954) would be shot in Natural Vision, although both were ultimately filmed in Warnerscope. Stereoscopic movies remained hot in the summer of 1953, when the fourth 3D feature, counting *Man in the Dark* mentioned in Note 4 above, appeared from Columbia. *Fort Ti*, directed by William Castle and photographed by Lester H. White and Lothrop Worth with a Natural Vision rig, was modestly budgeted compared to *House of Wax*, but it told an exciting story of the French and Indian War in matted widescreen (1.85:1, dubbed "Vistascope"),[40] Technicolor, and three-channel stereophonic sound. The film exploited a riot of emergence effects involving weapons hurled or fired from the screen, and it also featured the first 3D matte paintings. Accompanied by *Melody*, Disney's debut animated 3D short,[41] *Fort Ti* briefly became a box-office hit—in the first weeks of June, it was the nation's top-grossing film—but it was soon overshadowed by a plethora of more serious genre product (Figure 3.4). Universal's first 3D effort, for example, was *It Came from Outer Space* (Jack Arnold, 1953), based on a story by Ray Bradbury. To circumvent Natural Vision license fees, Universal authorized Jack Arnold and cameraman Cliff Stein to build their own, mirrorless dual-camera rig with a pair of Mitchells arrayed side-by-side, one of them mounted upside down, which was "a big monster," according to Arnold, but it worked. The film was released in Los Angeles on May 27 in "Wide-Vision" (matted 1.85:1) and three-channel stereophonic sound, tinted in what was advertised as "scientifically perfected eye-resting Full-Sepia Mono-Color!" (Word-of-mouth was already spreading that 3D viewing caused headaches and eyestrain, which could happen if projectionists failed to maintain the proper synchronization of their machines). This thrilling film—avowedly the inspiration for Steven Spielberg's *Close Encounters of the Third Kind* (1977)—remains one of the best, most realistic entries in the alien invasion subgenre. It was well reviewed and from late June through July replaced *Fort Ti* as the country's top-grossing hit. It also lent 3D a new credibility as a representational form—low on in-your-face emergence effects and high on the other-worldly, "supernatural" qualities of the process (Figure 3.5).

Paramount's first entry into 3D was *Sangaree* (Edward Ludwig, 1953), a big-budget costume film set on a plantation in 1780s Georgia, based on

---

40 *House of Wax* was frequently exhibited in the slightly cropped aspect ratio of 1.50:1, but restored to its original ration of 1.37:1 for the Warner Bros. 2013 Blu-ray reconstruction.

41 Distributed by RKO, the Disney short was poorly reviewed as having "cardboard cutout" 3D effects.

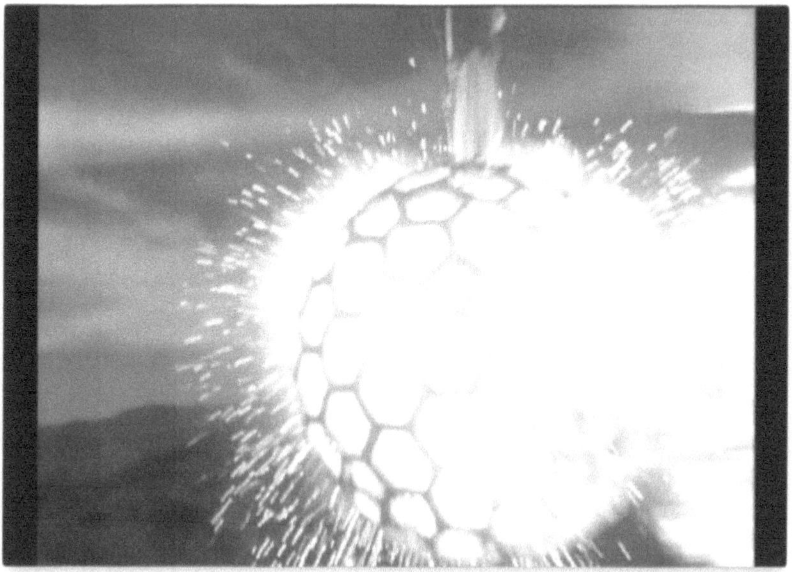

Figure 3.4 *It Came from Outer Space* (Universal Pictures, 1953): Universal's first 3D release, shot in black-and-white (advertised as "Full-Sepia Mono-Color") with an improvised Mitchell rig in "Wide-Vision" (matted 1.85:1), and three-channel stereophonic sound. Here, the meteorite hitting the audience in the face before the credit sequence.

a best-selling novel by Frank G. Slaughter and partially shot on location in Savannah. Like several of its predecessors, the production began "flat" and was switched to 3D after several major scenes had already been shot. Adolf Zukor, Paramount's CEO, ordered that the rest of the film be shot in a proprietary system called "Paravision." This was based on a camera designed in 1937 by Paramount's process photography expert, Farciot Edouard, for which he won the Academy's 1938 Scientific and Engineering Award. Then known as a "dual screen transparency camera," Edouard had developed it in order to film background plates that would double the room for studio action scenes by shooting two plates side-by-side and combining them horizontally in projection. The rig consisted of two 35mm cameras (originally Bell and Howells, then Mitchells) facing each other, with mirrors placed between them at a 45-degree angle, and it was ideal for stereoscopic photography. (In fact, Zukor claimed that the rig's 3D potential had been understood at the time, but "no one seemed interested in it until *Bwana Devil* made more money in its third week than *The Greatest Show on Earth* [a $4-million, star-studded C. B. DeMille

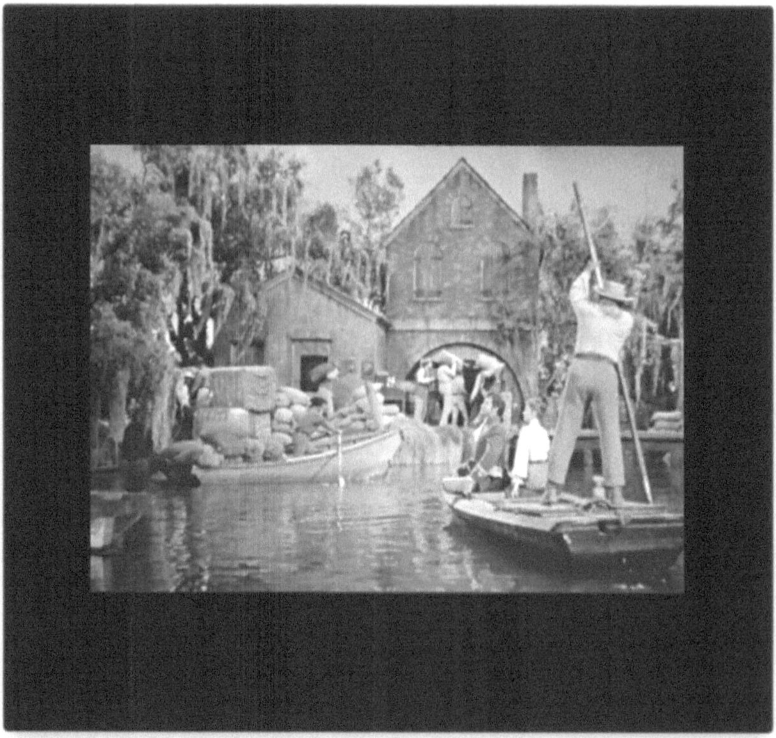

Figure 3.5 *Sangaree* (Paramount Pictures, 1953): Paramount's first 3D release, shot in "Paravision" and Technicolor, with a cropped aspect ratio of 1.66:1 and Western Electric magnetic stereophonic sound. Arlene Dahl and Fernando Lamas in perfect parallax.

blockbuster] made in its first week.")[42] Restarting production in "Paravision" and reshooting the "flat" scenes in 3D boosted *Sangaree*'s budget to $2 million, but Zukor's bet paid off: the film grossed $1.8 million domestically and another $3 million worldwide. Released in May in a copped aspect ratio of 1.66:1 with magnetic stereophonic sound, the film's reviews were positive, emphasizing its Technicolor cinematography, restrained use of 3D, and the charismatic appeal of its stars, Arlene Dahl and Fernando Lamas, who were romantically involved at the time (always good for publicity) and would later marry.

42 Cited in "Sangaree," http://www.3dfilmarchive.com. As noted in Chapter 2, Zukor also claimed that Paravision was based on the camera rig designed by Edwin S. Porter and William E. Waddle for the Famous Players Film Company's 1915 experimental screening of *Jim the Penman* and other shorts.

MGM produced its first stereoscopic feature, *Arena* (Richard Fleischer, 1953), using a proprietary rig called "Metro Tri-D," which like most others at the time involved two studio cameras mounted on a single lathe-like base. Shot partially on location in Tucson, Arizona, and released in May, this was a contemporary rodeo film—advertised as "the first 3-D western"—and it was released in "Variscope," with a cropped aspect ratio of 1.75:1 and magnetic stereophonic sound. Although it received mainly negative reviews, the plot of *Arena* is not unlike that of Sam Peckinpah's *Junior Bonner* (1971), concerning as it does an aging rodeo performer down on his luck who confronts a vicious Bramha bull, and it has the same motive force—dynamic rodeo acts as a showcase for cinematic spectacle (action in-depth here, stylized editing in Peckinpah). Audiences may not have liked the narrative, but they apparently liked the 3D because *Arena* grossed $1.2 million worldwide against a budget of $753,000, a healthy profit margin in 1953 (Figure 3.6).

The next important 3D production from an artistic point of view was Allied Artists' *The Maze* (William Cameron Menzies, 1953). Ironically, the roots of Allied Artists (AA) were in Monogram Pictures, a "Poverty Row" B-film factory founded in 1931. By 1953, the studio was grasping for respectability and discovered that it owned the rights to a 1945 novel by Maurice Sandoz, then among the leading Swiss writers of fantasy/horror, for which Salvador Dali had provided pen-and-ink illustrations. In February, AA announced plans to adapt the novel using Dali's designs in Natural Vision 3-Dimension. However, executive producer Walter Mirisch tried and failed to negotiate a licensing agreement with the Gunzbergs, whose leasing fees had skyrocketed thanks to doing business with major studios who were clamoring for their cameras. As was now happening in the industry on a regular basis, AA turned to one of its own camera engineers to construct a rival system, "Monogram 3-D," a twin-camera rig that used dual mirrors to achieve the necessary interaxial. On March 25, *Variety* announced that William Cameron Menzies had been chosen to direct—a major coup for AA, since Menzies was a Hollywood maestro who had won several Academy Awards for Art Direction (for *The Dove* [Roland West, 1927], *The Tempest* [Sam Taylor, 1928], and *Gone with the Wind* [Victor Fleming, 1939]) and who had directed several successful science fiction movies himself (*Things to Come* [1936] and *Invaders from Mars* [1953], among them). Dali was not directly involved in the production, but Menzies had collaborated with him on the dream sequence in Hitchcock's *Spellbound* (1945) and understood the surrealist's vision. The story is set in an old Scottish castle that contains an unspeakable family secret, and Menzies perfectly evoked a Gothic atmosphere reminiscent of Universal horror films of the 1930s and 1940s in terms of lighting and innovative set designs, including a perfectly constructed miniature maze. Released in 1.85:1 matted widescreen

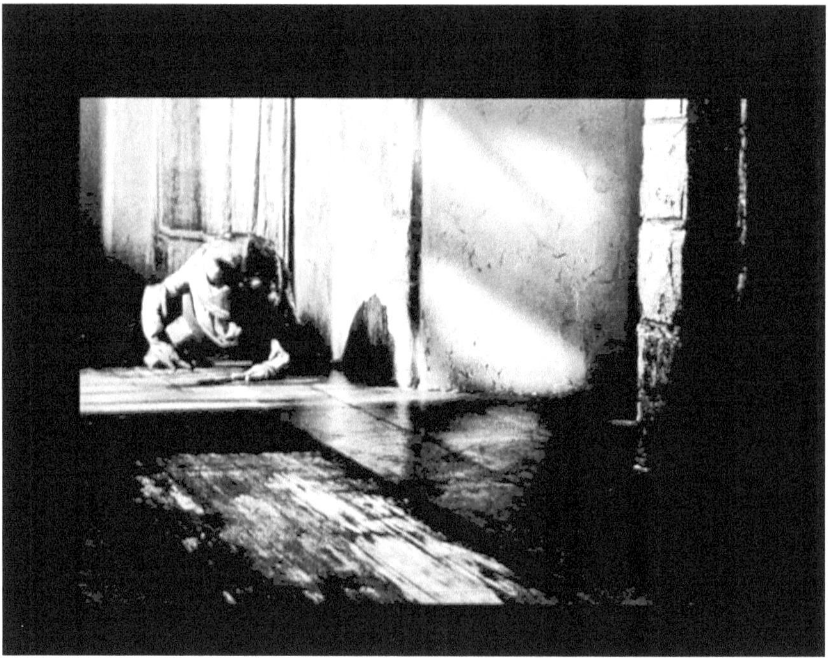

Figure 3.6 *The Maze* (Allied Artists Pictures Corporation, 1953): shot in "Monogram 3-D," black-and-white, matted to 1.85:1, with three-channel stereophonic sound by Western Electric. The monster leaves the maze and creeps into the castle.

with a three-channel magnetic stereophonic sound mix by Westrex, *The Maze* opened in 112 theaters nationwide over a five-week period during late June and July to great reviews and better business. The critic for *The Motion Picture Herald* wrote that "stereography is used in this finely developed presentation of a story [...] which explodes in the sharpest audience-response any 3-D film has elicited so far. Logical use is made of the depth medium to narrate a mystery [and] it builds to a terrific climax in which 3-D plays a vital part."[43] According to *Boxoffice*, "[*The Maze*] should determine whether the fast-growing and profit-assuring legion of 3-D fans are prepared to accept the process without the action that has characterized most earlier films that utilized it, because herein the chills and suspense are essayed through dialog and good performances, rather than movement and tempo."[44] With this encouragement, AA announced four upcoming widescreen 3D productions, but only one, *Dragonfly Squadron* (Leslie Selander, 1954), was actually shot in 3D and released

43 Ibid. "The Maze," http://www.3dfilmarchive.com. 10 October 2019.
44 Ibid.

flat, because studios had announced a "watch and wait policy" when public interest in the process suddenly took a nosedive.[45] As it turned out, *The Maze* was AA's only theatrically-released 3D film, and it was Menzies's last as a director, although it was artistically perhaps his most successful.[46]

The 3D boom, 1953–55

Between 1953 and 1955, Hollywood would go on to produce 69 features in 3D, and it is true that most of them were in action genres like the western, science fiction, and horror that could most effectively exploit the illusion of depth. A handful of these, such as *Robot Monster* (Phil Tucker, 1953), shot for less than $10,000, were laughably bad;[47] while others, like *Man in the Dark*, were short B-films released with great fanfare but suitable mainly to the bottom half of double bills. Yet there's no shame in exploiting the novelty of a new film technology—it's been done since the invention of motion pictures themselves (e.g., "the cinema of attractions") and the advent of sound, color, and widescreen. In June 1953, for example, Warner Bros. proved that depth could be exploited with style in *The Charge at Feather River* (Gordon Douglas), produced in Natural Vision, Technicolor, and WarnerPhonic sound. In this exciting Western, a cavalry troop rescues a white girl from a band of Indian abductors, and it is virtually built around emergence effects as arrows, tomahawks, spears, and bodies are hurled toward the audience from the screen during extended action sequences. Shot on location in Colorado and southern California and cropped to 1.66:1 for exhibition, *The Charge at Feather River* became the second highest-grossing film of the summer.[48]

---

45 *Dragonfly Squadron* was a melodramatic story set in the early years of the Korean War. It was produced in 3D, but given a general release in 2D in March 1954. It was restored to three dimensions on Blu-ray courtesy of Olive Film and the 3-D Film Archive in 2014.

46 According to biographer James Curtis, however, Menzies disliked the film and had limited control over its effects and script (Curtis, *William Cameron Menzies: The Shape of Films to Come* (New York: Pantheon Books, 2015), 327).

47 Added to this category must be two Columbia short produced in 3D for The 3 Stooges—*Pardon My Backfire* and *Spooks!* (both directed by Jules White, 1953). These unremittingly awful films are worth watching to see some of the worst excesses of 1950s stereoscopic cinema. Shots of characters throwing knives, meat cleavers, and flames (or squirting fluids) directly into the cameras lenses are cut into the narrative with little or no continuity matching.

48 It was also the last to use the WarnerPhonic mag/optical stereo system, composed of three magnetic tracks and one optical; after *The Charge at Feather River*, Warner's

July 1953 saw the release of RKO's first 3D feature, *Second Chance* (Rudolph Mate), a contemporary crime film shot with John A. Norling's Future Dimension camera system and released cropped at 1.66:1 in RKO "ScenicScope" and four- channel RCA magnetic sound. Norling's was a one-piece unit that ran two film strips behind two lenses, and it was lighter, quieter, and more mobile than other studios' two-camera rigs.[49] (RKO would eventually adopt the "SuperScope" format, which produced a standard ratio negative for optical conversion to anamorphic release prints in the lab.) (Figure 3.7).[50] Other 3D movies released in July–August 1953 were four westerns of variable quality—RKO's *Devil's Canyon* (Alfred Werker; Natural Vision, Technicolor, and ScenicScope), Columbia's *The Stranger Wore a Gun* (Columbia dual-strip 3-D, Technicolor, Vistascope), Universal's *Wings of the Hawk* (Budd Boetticher; Universal dual-strip 3-D, Technicolor, and Wide-Vision), and the independent Realart's *Hannah Lee: An American Primitive* (Lee Garmes and John Ireland; Stereo-Cine dual-strip 3D, Pathecolor, Academy ratio)—and two rather impressive crime films. These latter were United Artists' *I, The Jury* (Harry Essex; StereoVision dual-strip 3-D, black-and-white, cropped to 1.75:1), based on Mickey Spillane's best-selling Mike Hammer novel, with cinematography by *noir* master John Alton, and Fox's *Inferno* (Roy Ward Baker; Clear-Vision dual-strip 3-D, Technicolor, cropped to 1.66:1), a film noir in color with an all-star cast in which a wealthy businessman (Robert Ryan) is lured into the Mohave Desert by his cheating wife (Rhonda Fleming) and her lover, injured, and left for dead in a scheme to inherit his fortune. His struggle to survive the wasteland with a broken leg and seek revenge was beautifully photographed on location by veteran DP Lucien Ballard (Figure 3.8).

In the summer of 1953, only about half of the theaters in the United States were equipped to show 3D; the cost of installing an extra projector was about $2,000 (although most had two already, and a package of two interlocking motors to synchronize them could be had for $710),[51] and many exhibitors

---

converted to the magnetic four-channel format used by the other studios except Paramount (Hayes, *3-D Movies*, 29).

49 Hayes, *3-D Movies*, 29.

50 When the 1950s boom ended, RKO converted a number of its unreleased 3D titles to SuperScope, compromising their 1.37:1 aspect ratios by cropping them to 1.66:1 or even 2.00:1. An example is *Son of Sinbad* (Ted Tetzlaff, 1955), originally shot in "ScenicScope" and scheduled for 3D release in 1953 but shelved due to Production Code interference. (Joseph Breen objected to erotic belly dances by the scantily clad Lili St. Cyr, a well-known strip-tease artist of the time; these were cut from the credit sequence and reedited in the film itself.) Released in anamorphic 2.00:1, the top and bottom of the frame were cut off, destroying the careful composition of many shots.

51 Zone, *3D Revolution*, 73.

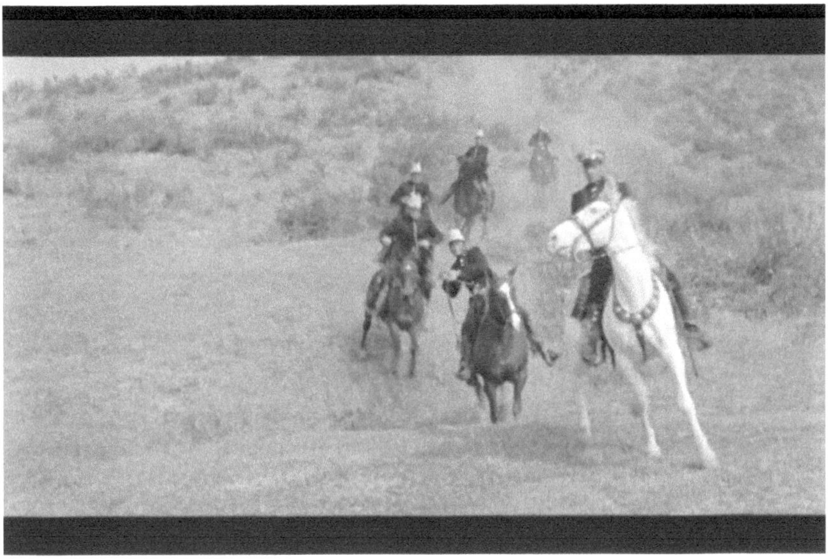

Figure 3.7 *Wings of the Hawk* (Universal Pictures, 1953): shot with Universal's Mitchell rig (now "Universal 3-D") in Technicolor, "Wide-Vision" (matted 1.85:1), and Western Electric stereophonic sound. Emperor Maximillian's cavalry in pursuit of Mexican gold.

were waiting to observe the public reception of CinemaScope. After *The Robe* premiered in the new process on September 16 at the Roxy Theater in New York to critical raves and took in an astonishing $270,000 in its first week, Hollywood's ardor for stereoscopy began to cool and theater-owners grew less willing to add the new projection and sound equipment required for it. In the following week, Warner Bros. released the black-and-white 3D western *The Moonlighter* (Roy Rowland), teaming Barbara Stanwyck and Fred McMurray as lovers again for the first time since *Double Indemnity* (Billy Wilder, 1944), which bombed at the box office. Soon, Paramount would become the first studio to offer films to exhibitors in both 3D and 2D versions. Before that, 3D films had been available in flat format only after they had played all of their 3D engagements (Figure 3.9). *Those Redheads from Seattle* (Lewis R. Foster; Paravision, Technicolor, Panoramic Screen [1.66:1]), advertised as "The First 3-D Musical," was released in both formats in mid-October, followed by the espionage thriller *Flight to Tangier* (Charles Marquis Warren; Dynoptic dual-strip 3-D, Technicolor, Panoramic Screen) in November.[52] *Redheads* was the

52 Apparently, *Flight to Tangier* and Paramount's Dean Martin/Jerry Lewis comedy *Money from Home* (George Marshall; Dynoptic 3-D, Technicolor, Panoramic Screen), released

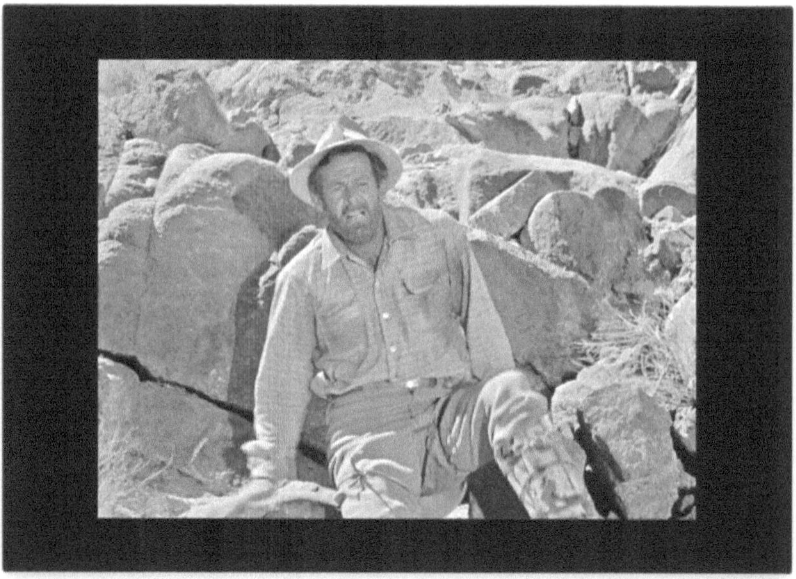

Figure 3.8 *Inferno* (20th Century Fox, 1953): shot in Stereo-Scopic Clear-Vision dual-strip 3-D, Technicolor (cropped to 1.66:1), and Western Electric stereophonic sound. Robert Ryan abandoned in the Mojave Desert.

first film actually composed for a widescreen ratio, rather than cropped for one after the fact, but its dual-format release seemed to confirm the waning popularity of 3D: of its 16,464 domestic bookings, just 345 were for the 3D version.[53] The only other 3D release of the month was Universal's *The Glass Web* (Jack Arnold), a modest black-and-white film noir starring Edward G. Robinson (Figure 3.10).

In November, MGM decided to premiere its big-budget adaptation of Cole Porter's long-running Broadway musical *Kiss Me Kate* (George Sidney) in both 3D and flat versions. Like the play, the film revolves around the staging of a musical version of Shakespeare's *The Taming of the Shrew* in which formerly married partners play the leads; it was written, shot, and choreographed (mainly by Hermes Pan) specifically for 3D, but MGM wanted to test the

in December, were the only 3D films shot in 3-strip Technicolor, thus requiring six cameras.

53 Quoted in Hillary Hess, Greg Kintz, Jack Theakston, and Bob Furmanek's Audio Commentary for *Those Redheads from Seattle* Blu-ray (Kino Lorber/3-D Film Archive, 2017).

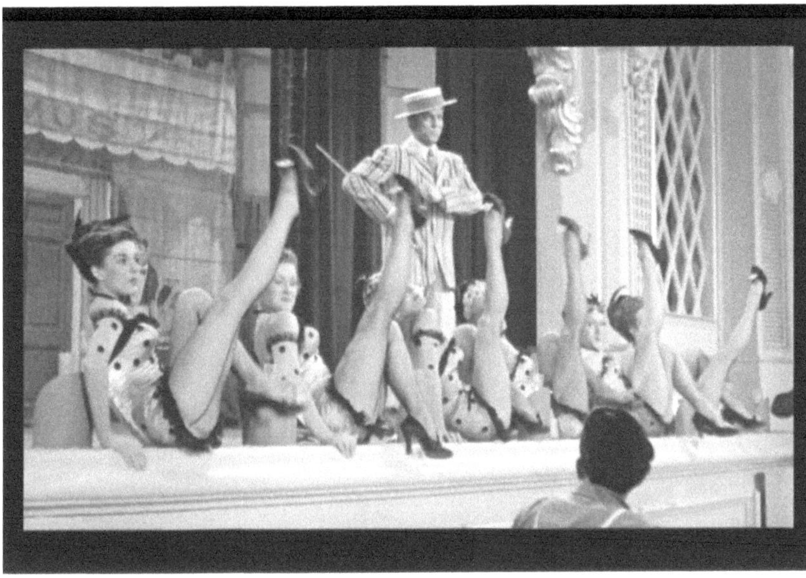

Figure 3.9 *Those Redheads from Seattle* (Paramount Pictures, 1953): shot in Paravision, Technicolor, and "Panoramic Screen" (1.66:1), with Western Electric stereophonic sound, advertised as "The First 3-D Musical." It was also the first film actually composed for a widescreen ratio, rather than cropped for one after the fact, and the first to be released both in 2D ("flat") and 3D versions. A chorus line.

public's wavering enthusiasm for stereoscopy through the dual release. The results were a false positive: the 3D version did nearly twice the business of the 2D version, leading MGM's head of sales to announce, "In light of these experiments, we strongly urge theatres to play *Kiss Me Kate* in 3-D."[54] Yet the film lost money domestically against its $2 million cost,[55] and studio boss Dore Schary declared that it would be MGM's last 3D production (Figure 3.11) The very next week, Warner Bros. released its fourth 3D film, *Hondo* (John Farrow), shot with a new rig designed by its own camera department—the Warner All-Media dual-lens camera. It used a beam-splitting prism that could be adjusted to provide a very small separation between its lenses for close-ups without causing background divergence—a big problem for the fixed-lens Natural Vision system whose interaxial distance was 3.5 inches or, as stated earlier, one inch greater than the probable distance between

54 Quoted in Morgan and Symmes, *Amazing 3-D*, 86.

55 *Kiss Me Kate* was made for $1,981,000 and it grossed $3,117,000 worldwide, producing a loss of $544,000 after deducting for distribution and other costs.

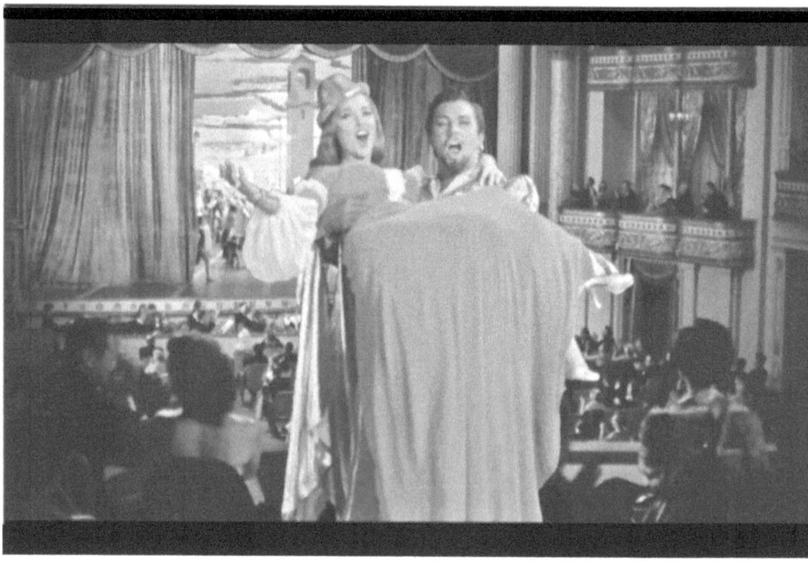

Figure 3.10 *Kiss Me Kate* (Metro-Goldwyn-Mayer, 1953): shot in Metrovision Tri-Dee, and presented in Variscope (1.75:1) and Western Electric stereophonic sound. Stars Kathryn Grayson and Howard Keel emerging from both stage and screen into audience space at the end of both play and film.

the human eyes. Branded as WarnerVision 3-D, the new process used a smaller rig than any of the twin-camera processes, though still bulky, and it was especially good for close-quarter filming in a relatively enclosed space, such as the small set that would be used in Hitchcock's *Dial M for Murder* (1954). *Hondo* was based on a Louis L'Amour story and produced for Warner Bros. distribution by the new John Wayne/Robert Fellows Batjac Productions company as its fourth feature; it was shot on location by DP Robert Burks in the state of Chuhauhau, Mexico, and it emphasized the depth perspective of its desert vistas rather than emergence effects, although it did contain some notable "pop-outs" in action sequences. (*Hondo*'s exciting conclusion featured an Apache attack on a wagon train that was directed, uncredited, by John Ford, when Farrow had to leave the set to begin another project.) Warner released the film in 3D in major cities, cropped for a widescreen ratio of 1.66:1, but smaller theaters screened it in 2D; nevertheless it was popular with audiences in both versions, grossing $4.1 million to become one of the top-earning films of the year and the second most profitable feature of the 1950s 3D boom after *House of Wax* (Figure 3.12).

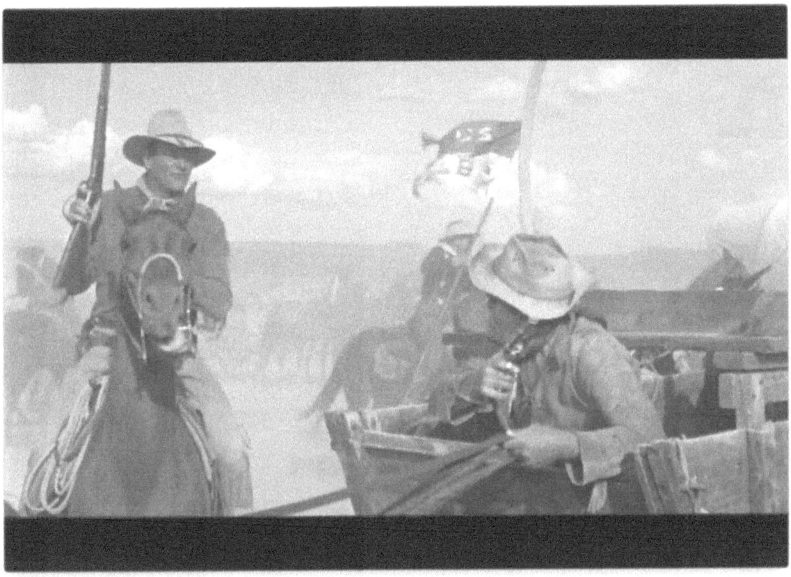

Figure 3.11 *Hondo* (Warner Bros., 1953): shot in reengineered Natural Vision (branded as "WarnerVision 3-Dimension"), in Technicolor (branded as "WarnerColor"), Widescreen (cropped to 1.66:1), and RCA Photophone stereophonic sound. John Wayne and Ward Bond in the final battle sequence.

Also released in November was Columbia's equally well-produced western *Gun Fury* (Raoul Walsh), shot with a 3D rig that improved upon the makeshift one assembled for *Man in the Dark* seven months earlier. Similar to Universal's two-camera process, the "Columbia 3-D" system employed dual Mitchell camera movements mounted side-by-side, one of them turned upside down, and shooting straight ahead without the intervention of mirrors or prisms As noted by Ray Zone, "Though housed within a blimp for noiseless filming, the Columbia stereo camera was a smaller mobile unit that provided control of convergence and variable interaxial operated by knobs and dials on either side of the blimp."[56] The rig also had follow focus controls for moving camera sequences, and its ability to narrow the distance between its lenses facilitated sharp focus across a range of shot scales, from close-ups to deep backgrounds. Filmed on location in the Red Rocks area of Sedona, Arizona, by DP Lester White, *Gun Fury* offered compositionally beautiful stereoscopic realism in the

56 Zone, *3D Revolution*, 37.

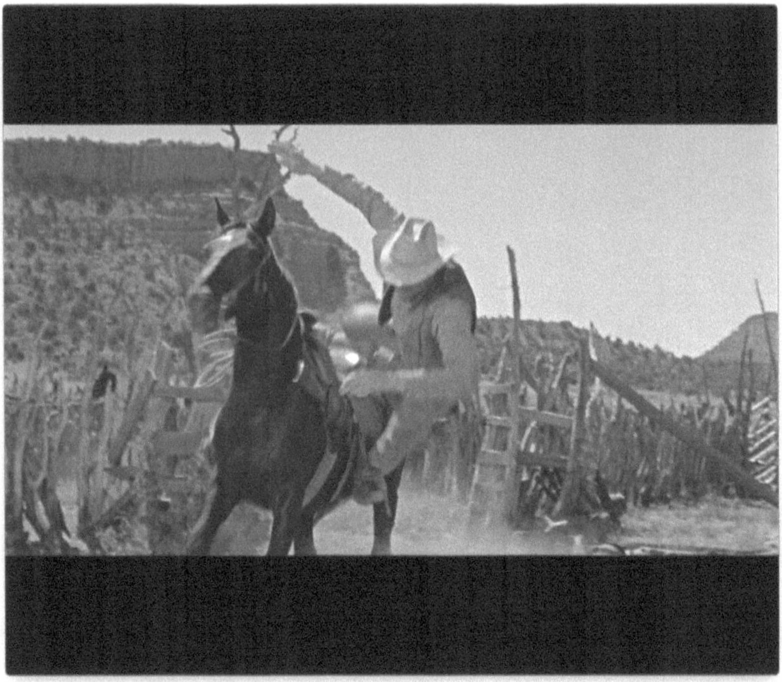

Figure 3.12 *Gun Fury* (Columbia Pictures, 1953): shot in "Columbia 3-D" and presented in "Nord 3-D" (single-strip side-by-side vertical 1.75:1), Technicolor, and RCA Photophone stereophonic sound. One of many inventive negative parallax effects: a rider spills into audience space.

tale of a young woman (Donna Reed) betrayed and abducted by an outlaw gang with her fiancée (Rock Hudson) in hot pursuit (Figure 3.13).

The last notable 3D film of 1953 was Columbia's pre-Christmas release on December 23 of *Miss Sadie Thompson* (Curtis Bernhardt). Based on the Somerset Maugham short story "Rain" adapted twice before,[57] it was originally conceived as a big-budget Technicolor musical highlighting the dancing talents of Rita Hayworth, but only three musical numbers actually made it into the film, and Hayworth's singing was dubbed in all of them (by Jo Ann Greer). Exteriors for

57 As *Sadie Thompson* (Raoul Walsh, 1928), a much-acclaimed Gloria Swanson vehicle, and, less successfully, as *Rain* (Lewis Milestone, 1932), starring Joan Crawford. Both films were controversial in their day for their depiction of a "fallen woman," as was the race film *Dirty Gertie from Harlem U.S.A.* (Spencer Williams, 1946), an unauthorized adaptation.

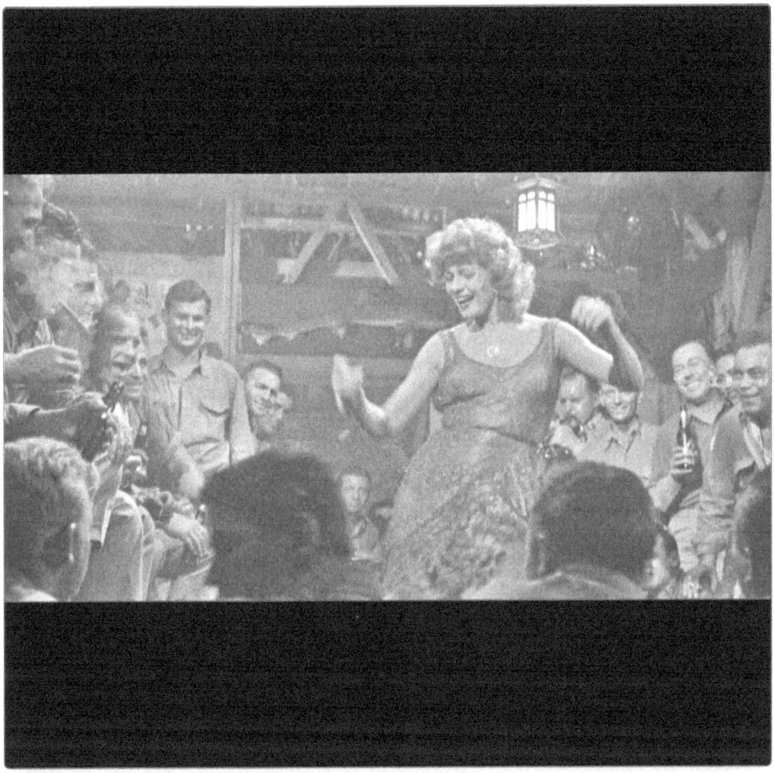

Figure 3.13 *Miss Sadie Thompson* (Columbia Pictures, 1953): shot in "Columbia 3-D" and presented in "Nord 3-D" (single-strip side-by-side vertical 1.75:1), Technicolor, and RCA Photophone stereophonic sound. Rita Hayworth as Sadie sings "The Heat Is On."

Miss Sadie Thompson were shot on location in Kauai, Hawaii, and interiors on the Columbia lot, with an emphasis on dramatic involvement through depth rather than emergence effects. The film was notable for being censored in several Deep South states for the sensuousness of its dance routines (especially "The Heat Is On") and the fact that the Thompson character—despite being labeled a "dance-hall girl" by the script to appease the Hays Office—is very clearly a prostitute, as in the original story and its other film adaptations. The film was shot in Columbia 3-D; prints were cropped at 1.75:1 and featured three-channel RCA stereophonic sound. After a two-week opening, Columbia withdrew the 3D version and continued to distribute the film flat. *Miss Sadie Thompson* grossed $2.9 million to return a modest profit.

The new year began inauspiciously with the independent Astor's *Cat-Women of the Moon* (Arthur Hilton) and Columbia's *Drums of Tahiti* (William Castle), both short, reasonably efficient programmers. In January, however, Paramount released a most unusual 3D film, shot in Paravision, produced by Warner Bros. veteran Hal B. Wallis (*Casablanca, The Maltese Falcon*) and directed by Owen Crump, who had supervised production for the Signal Corps' First Motion Picture Unit in Culver City during the Second World War. Filmed on actual battlefield locations, *Cease Fire* cast real soldiers instead of professional actors in a story of an infantry platoon in combat on the last day of the Korean War, which an introduction by General Mark W. Clark, former commander-in-chief of the United Nations forces, describes as an "accurate and authentic" depiction of the conflict (Figure 3.14). Despite its low-budget black-and-white imagery, the film's use of stereoscopy is impressive, and its opening shot is what Martin Liebman calls "perhaps the most distinctly unique ever seen in 3D, with the elongated barrel [of a long cannon] protruding from the screen while it's simultaneously easy to look far deep down into it."[58]

That 3D was continuing to evolve technically was demonstrated by two Universal releases in February. Like *Miss Sadie Thompson, Taza, Son of Cochise* (Douglas Sirk) was shot in Pola-Lite, a single-strip side-by-side vertical process, and released in both dual-strip Vistascope and "Nord 3-D," a single-strip vertical projection format with an aspect ratio of 1.75:1 that exhibitors could show using only one projector. A sequel to Fox's popular *Broken Arrow* (Delmer Daves, 1950), *Taza* was shot on location in Utah and boasted spectacular Technicolor photography by Russell Metty (Figure 3.15). More innovative was *Creature from the Black Lagoon* (Jack Arnold), a horror film about a Gill-man in the Amazon River that attacks a scientific party attempting to study it, whose script required multiple underwater sequences. Clifford Stine, a special effects photographer in Universal's camera department, devised an underwater 3D rig for *Creature* that consisted of two small 35mm Arriflex cameras mounted in a compact, watertight housing, radical for the time since such cameras were deemed too flimsy for professional use. The mesmerizing underwater sequences were filmed at Wakulla Springs, Florida, and featured Olympic swimmer Rico Browning in costume as the Creature underwater (with Ben Chapman as the Creature on land) (Figure 3.16). The black-and-white film proved extremely popular and generated two sequels—one of them, *Revenge of the Creature* (Jack Arnold, 1955), was the last film of the 1950s 3D boom, while the other, *The Creature Walks among Us* (John Sherwood, 1956), was shot and

58 Marc Liebman, "*Cease Fire* 3D Blu-ray Review," www.blueray.com, March 29, 2018.

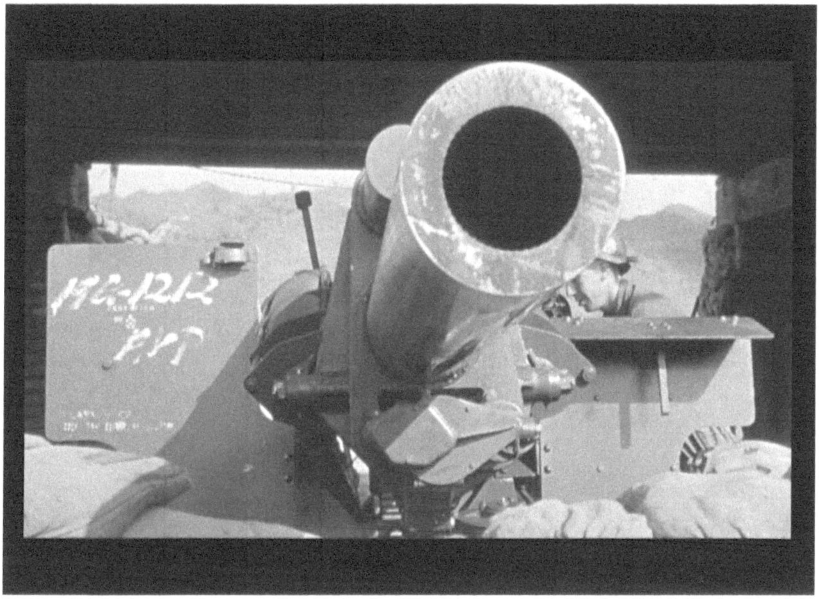

Figure 3.14 *Cease Fire* (Warner Bros., 1954): shot in Paravision and "Panoramic Screen" (cropped 1.85:1), in black-and-white, with Western Electric stereophonic sound. A cannon barrel in negative parallax.

released flat. *Creature* was also the inspiration for Guillermo del Toro's award-winning 2017 film *The Shape of Water*.

Like *Taza, Son of Cochise*, *Creature from the Black Lagoon* was released in Nord 3-D so that theaters could show it using a single projector. As audience enthusiasm for 3D began to decline, exhibitors were less and less inclined to install the extra equipment required by the format, and several single-strip projection systems found favor in 1954—all of them combining the two camera shots on a single film strip and projecting the alternate frames through a device that cross-polarized and superimposed them on-screen. Nord introduced the first such system via Columbia, and other studios subsequently adopted it. Another system, invented by Boris Morros, was called Pola-Lite, and it competed briefly with Nord, but, as R. M. Hayes points out, "While the single-strip formats eliminated mis-synchronization, they also suffered from a great deal of light loss," and, in any case, by the time they became available "the situation was really beyond saving."[59] What Hayes meant was that by the early spring of 1954, the 3D boom was nearly over.

59 Hayes, *3-D Movies*, 41.

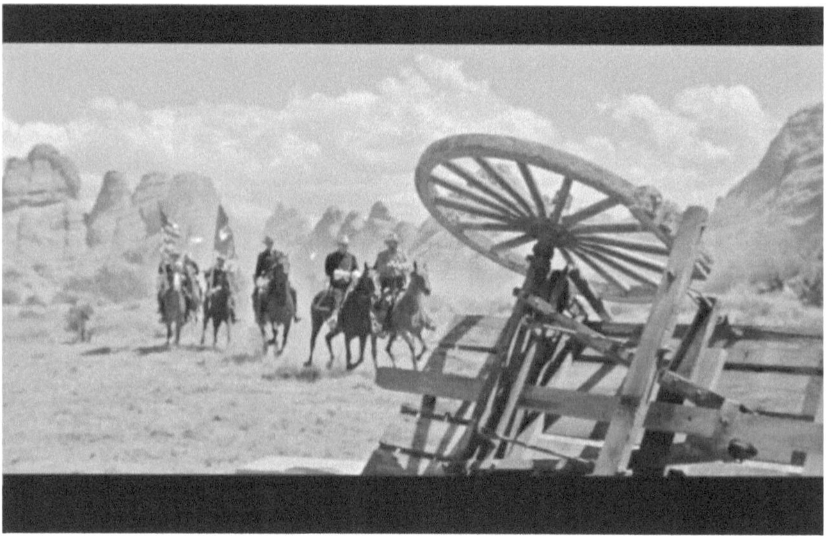

Figure 3.15 *Taza, Son of Cochise* (Universal Pictures, 1954): shot in Universal 3-D and presented in both Pola-Lite (single-strip side-by-side 1.66:1) and "Nord 3-D" (1.75:1), with Technicolor and Western Electric stereophonic sound. This impressive shot shows perfect parallax in its continuity of depth and emergence.

Yet in some quarters the fires were still burning. In February, RKO released the Jane Russell musical *The French Line* (Lloyd Bacon), which flaunted its star's skimpy costumes and fetishized the stereoscopic presentation of her breasts ("See J.R. in 3-D." the ads read. "It'll knock *both* your eyes out!") The Production Code Administration denied the film its seal of approval, branding it "offensive" and citing Russell's "Gettin' into Trouble" dance sequence for "indecent exposure." RKO was fined $25,000 for distributing the film nationwide, and the dance number in question was censored or cut in many cities. Where it wasn't banned outright (as it was, e.g., in Chicago), *The French Line* did fantastic business, and, despite relatively low production values, ended up grossing nearly $3 million. That same month Warner's released an "official" follow-up to *House of Wax*, *Phantom of the Rue Morgue* (Roy Del Ruth). Based on the Poe short story "Murders in the Rue Morgue" and first adapted under that title by Robert Florey in 1932, this film offered excellent three-dimensional composition in the cinematography of J. Peverell Marley and broke even with a domestic gross of nearly $1.5 million. In May 1954, Columbia showed that the exploitation of depth could still be done with taste in *The Mad Magician*, another sequel, of sorts, to *House of Wax*, featuring some of the same cast and creative personnel, most notably Vincent Price as the titular lead (Figure 3.17). Hastily produced but stylishly directed by John Brahm, whose successful

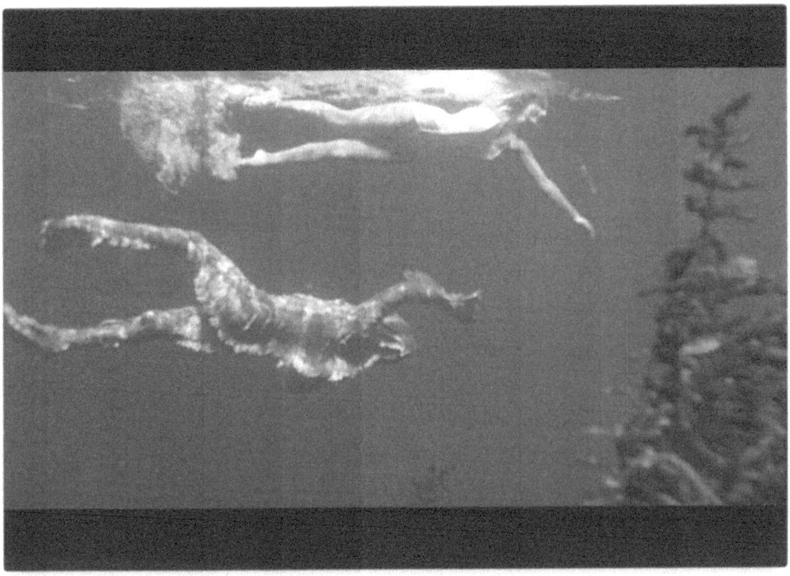

Figure 3.16 *Creature from the Black Lagoon* (Universal Pictures, 1954): shot in Universal 3-D in black-and-white and presented in Pola-Lite (1.66:1) and Western Electric stereophonic sound. Underwater ballet with Julia Adams (unaware) and the Creature; underwater photography by Charles Scott Welbourne.

gothic thrillers *The Lodger* (1944) and *Hangover Square* (1945) had their roots in German Expressionism, the film made excellent use of layered 3D in its art direction and atmospheric black-and-white photography by Bert Glennon (who had shot many of John Ford's best films, including *Stagecoach* in 1939). Also in May, Fox released *Gorilla at Large* (Harmon Jones) in Pola-Lite, the single-strip vertical side-by-side process noted above. (Although the studio was committed to CinemaScope, it had formed a partnership with Leonard Goldstein's Panoramic Productions to make non-anamorphic films, of which *Gorilla at Large* was the first.) A mystery-horror film with a low budget but a great cast (e.g., Anne Bancroft, Lee J. Cobb), it was shot on location at the Nu Pike Amusement Park in Long Beach and featured inventive 3D Technicolor cinematography by Lloyd Ahern; unfortunately, it was revived in a network television gimmick in the early 1980s as part of a special promotion with 7/11 convenience stores and Revco pharmacies, which provided cardboard glasses that failed to synchronize with most receivers.

In May 1954, too, Warner's released the last and best of its boom features in three dimensions. Expertly shot in WarnerVision and WarnerColor by Robert

Burks, Alfred Hitchcock's *Dial M for Murder* is reference-quality stereoscopic cinema—quite literally, since both Martin Scorsese and Baz Luhrmann screened it to prepare their crews for shooting *Hugo* (2011) and *The Great Gatsby* (2013), respectively, in digital 3D (see Chapter 7). Both the screenplay and long-running stage play on which it was based were written by the English scenarist Frederick Knott, and the movie is filmed theater of the highest order. In fact, *Dial M for Murder* (1954) is one of the greatest films ever made in 3D, and yet it barely exploits the process at all. Only when Grace Kelly reaches out into audience space to grasp the scissors with which she kills her murderous intruder in self-defense does a truly dynamic emergence effect appear. It is the functional equivalent of Marion Crane reaching out to the audience for empathy at the end of the shower murder sequence in *Psycho* (1960), except that here the fourth wall is broken not figuratively but literally, and the identification it creates with Kelly's character is stunning. (Another striking emergence shot

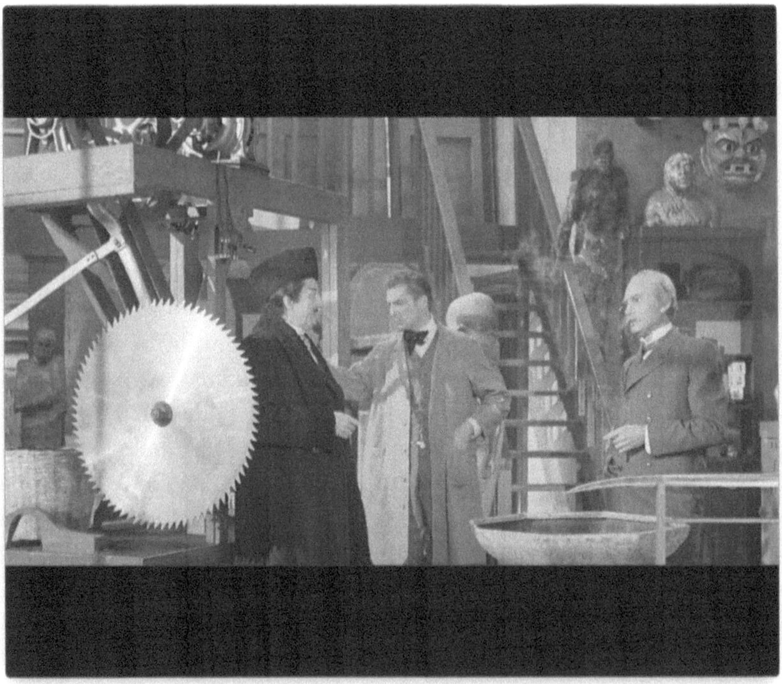

Figure 3.17 *The Mad Magician* (Columbia Pictures, 1954): shot in Columbia 3-D in black-and-white and presented in Nord 3-D (1.75:1) and RCA Photophone stereophonic sound. The buzz in negative parallax and Vincent Price as inventor and mad magician, center-frame.

occurs near the end of *Dial M* when the inspector extends the latchkey in his hand as the solution to the mystery.) (Figure 3.18). There has been a persistent rumor that Hitchcock did not plan to produce the film in WarnerVision, but he actually story-boarded it for three dimensions from the outset, including the use of some notably flat rear-projection for backgrounds, and composed it in a ratio of 1.85:1. Throughout his career the director had eagerly embraced technical innovation—the Schufftan process (for special effects) and sound in the late 1920s and 1930s, color in the 1940s, and widescreen in the 1950s—and stereoscopy was no exception. As Bob Furmanek and Greg Kintz have written of *Dial M for Murder*, "Every shot was composed, so that with proper camera movement and very precise convergence, virtually error-free 3-D was achieved."[60] The film premiered in 29 major cities and was aggressively marketed as "Perfect 3 Dimension in the Hitchcock Manner." It was also well reviewed as such, *The Hollywood Reporter* calling it "one of the few films in which 3-D is a decided asset," and *Time* noting that "The 3-D is used not so much for its shock value as to bring alive for moviegoers much of the theatre's intimacy and depth of movement."[61] Unfortunately, audiences had begun to reject the process because of flawed presentations, and Warner dropped the 3D version, which effectively disappeared until a failed theatrical revival in the 1980s and a superb Blu-ray restoration in 2012. Most audiences in 1954 saw *Dial M for Murder* flat, and it became a hit in that format; Jack Warner, 3D's strongest industry proponent, subsequently gave up on it, and Warner's last three-dimensional feature, a Randolph Scott western called *The Bounty Hunter* (Andre de Toth, 1954), was released in 2D only in September.

The only 3D feature released in July 1954 was United Artists' science fiction film *Gog*, shot in SuperCinecolor,[62] Natural Vision, and widescreen (1.66:1) by Lothrop Worth, and directed and edited by Herbert L. Strock. The third film in producer Ivan Tors's "Office of Scientific Investigation" (OSI) trilogy, which included UA's 2D *The Magnetic Monster* (Curt Siodmak, 1953) and *Riders to the Stars* (Richard Carlson, 1954), *Gog* was carefully planned for 3D presentation. Shot in just 15 days on a budget of $250,000 in September 1953, the film nevertheless features excellent three-dimensional cinematography. It concerns the sabotage of a super-computer that controls America's nuclear defense system by an unspecified enemy and climaxes in a murderous rampage by

60 Bob Furmanek and Greg Kintz, "An In-Depth Look at *Dial M for Murder*," http://www.3dfilmarchive.com.

61 Quoted in ibid.

62 A three-color process similar to Anscocolor that briefly produced color prints for just 10 percent more than black-and-white.

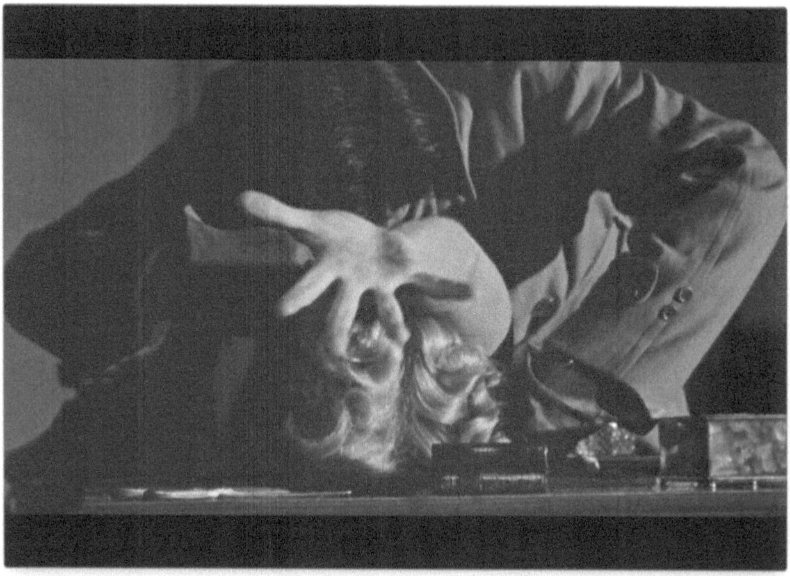

Figure 3.18 *Dial M for Murder* (Warner Bros., 1954): shot in WarnerVision 3-D and Widescreen (cropped 1.85:1) in WarnerColor and RCA Photophone stereophonic sound. Grace Kelley's hand breaks the fourth wall in precise convergence, achieving "virtually error-free 3-D."

two large robots, Gog and Magog, named after Old Testament entities. Among *Gog*'s salient points is that art director Lyle Wheeler, a multiple Oscar winner, did research into the psychology of color to increase its emotional impact and the fact that its top-secret underground laboratory provided the inspiration for Michael Crichton's 1969 novel *The Andromeda Strain*, adapted to film two years later by Robert Wise. Furthermore, director Herbert Strock had monocular vision—like Andre de Toth (*House of Wax*) and Raoul Walsh (*Gun Fury*)—making him the third notable 3D director who could not see normally in three dimensions, so he relied on Natural Vision pioneers Milton and Julian Gunzberg as visual consultants to help him set up his shots to their best stereoscopic effect (Figure 3.19). Although it was given a wide release, only five theaters, all in Southern California, played *Gog* in 3D because the fad was dying. Ironically, as the quality of 3D production increased, fewer and fewer audiences were able (or wanted) to experience films in that format.

By the late summer and early fall of 1954, nearly all features that had been shot in 3D were released flat, producing a double inventory of prints (one for

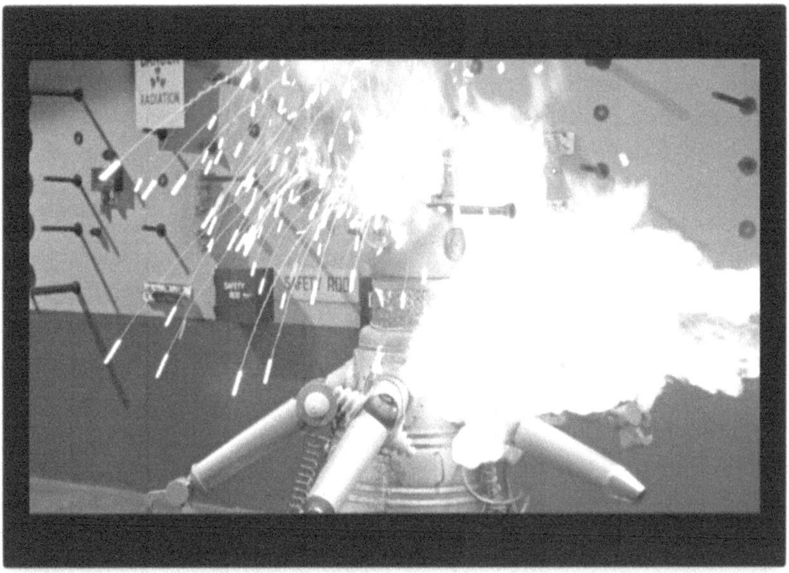

Figure 3.19 *Gog* (United Artists Corporation, 1954): shot in Natural Vision and presented in Pola-Lite (1.66:1) and SuperCinecolor (a variation of Anscocolor), with Western Electric stereophonic sound. Gog incinerated.

each eye, naturally, and each slightly off-center in terms of composition). The last 1950s film produced and exhibited in the process was *Revenge of the Creature* (Jack Arnold, 1955), a direct sequel to *Creature from the Black Lagoon* shot in Universal 3-D (dual 35mm) and presented in "3-D Horrorscope," a brand name for Pola-lite single-strip side-by-side vertical composed at 1.66:1. The film fine-tuned the Creature costume and gave him a kind of rudimentary psychology; it was shot on location by Charles Scott Welbourne, with Marineland of Florida standing in for the "Ocean Harbor Oceanarium," where the Gill-man has been confined for study by ichthyologists. The underwater sequences once again featured Ric Browning as the Creature and were shot in black-and-white by Jack Lamb, who would become the cameraman for the popular Lloyd Bridge syndicated television series "Sea Hunt" (1958–61). *Revenge of the Creature* retained several characters from its predecessor and contained some impressive three-dimensional sequences; despite its late-stage appearance, the film was a modest commercial success in both its 3D and flat versions. It was also the last stereoscopic film to be released commercially in the United States until 1960.

The end [...]

Speculation about reasons for the sudden decline of 1950s 3D has often centered on the quality of the films themselves and the discomfort caused by 3D eyewear. In fact, the situation was considerably more complex. The main reason behind audience disillusionment with the process was neither the films nor the glasses, but poorly executed 3D projection. Vertical misalignment of the two interlocked projectors was common and could be caused by the shredding of a single sprocket hole. Such tears could cause havoc with optical convergence, producing blurred and headache-inducing images for spectators as one eye struggled to keep up with the other.[63] As Jack Theakston has written, "Essentially, the sloppy presentations and inability to maintain synchronization ***is*** the reason why 3-D failed."[64]

Another problem involved the practical cost of 3D to exhibitors. While large urban theaters seating thousands could amortize the cost of new equipment, smaller venues, on which the industry depended for its nationwide distribution network, could not. As R. M. Hayes points out, increased color production, matted widescreen, CinemaScope, stereophonic sound, and 3D had all been set forth as an antidote to television; all cost millions of dollars to develop and provide, and the studios passed those costs along to exhibitors. The expense of installing new projectors, screens, lenses, and sound systems was steep, yet theater chains have always been at the mercy of Hollywood's supply chain—with no product to screen, an exhibitor owns nothing but an empty building.

---

63 The issue of headaches has never been fully resolved, and audiences continued to complain of them into the eras of single-strip and digital 3D when the left- and right-eye images *could technically not* get out of synchronization. According to www.medicaldaily.com, a study presented by Dmitry Vatolin, senior research fellow in Graphics and Media Lab, Lomonosov Moscow University, at the 27th annual conference Stereoscopic Displays & Applications in San Francisco in February 2016 revealed that over 10,000 problematic scenes from 105 different 3D movies were shot with the "wrong" images hitting both eyes simultaneously, often by a reordering of the left and right views. After eight years of research, Vatolin and his team determined 15 reasons for headaches caused by watching 3D movies and split them into two categories—equipment imperfections and production errors. Imperfections of equipment, they found, were often a function of low budgets and inferior camera systems, especially in genres like horror, but the 3D movie viewing experience could also be ruined by a bad pair of glasses or cheap projectors. Production errors, on the other hand, could occur in high-budget movies as well, when the swapping of scenes from left to right sent scrambled signals to the brain, which Vatolin found in about 21 percent of the movies in his sample. Vatolin did not explain how and why this swapping occurs, but the problem it produces is clearly "parallax distortion" as described in the subsection "Parallax, Convergence, and Focal Length" (pp. 40–42).

64 Jack Theakston, "What Killed 3-D?," http://www.3dfilmarchive.com.

Hayes puts it succinctly: "Hollywood was at war with television, and as a result new technologies had been developed and were going to be used across the board. […] Exhibitors had no choice but to go widescreen and CinemaScope […] [but] they did have a choice regarding 3-D and stereophonic sound, and they decided to take the less expensive route."[65]

Although widely advertised as "3-D You Can See without Glasses," CinemaScope provided nothing like the illusion of depth created through stereoscopy. At best, its wide-angle anamorphic lenses and deeply curved screen enhanced spectator involvement, as did its four-track magnetic sound, but in terms of sustaining a true depth perspective it wasn't even poor man's three-dimensional cinema. (In fact, early anamorphic lenses tended to distort backgrounds and middle grounds in close-ups, although subsequent refinements gave them greater depth of field.)

The triumph of CinemaScope led to a succession of wide-gauge film systems, most of which used some version of 70mm negative stock (65mm for the picture, 5mm for a magnetic stereophonic soundtrack) to produce vivid, highly resolved widescreen images. The equipment was big and bulky—necessarily twice the size of standard 35mm—but no more so than 3D rigs, and it produced twice the picture information on-screen as its anamorphic cousins. (The anamorphic image cast on a 60-by-30-foot screen lost clarity and brightness because it was distributed across too large a field through decompression.) The first and most unique of them was Paramount's VistaVision, which ran a 35mm negative through a horizontal-feed camera to expose two frames at a time, producing an image that was twice as wide as conventional 35mm and slightly taller; the result could be shown full-frame on horizontal transport projectors to produce gigantic 1.96:1 images (as it was on its premiere in Michael Curtiz's *White Christmas* at Radio City Music Hall in October 1954) or rotated optically and printed down to 35mm for vertical projection at 1.85:1 in nonspecialized theaters. Unlike other wide-film processes, VistaVision did not employ stereomagnetic sound but rather a single-track optical system called Perspecta sound that could be combined with a cost-effective processor to produce simulated three-channel stereo.

Next came the 70mm wide-film system called Todd-AO, developed by independent producer Mike Todd and the American Optical Company; it was designed to compete not only with CinemaScope but with Cinerama as well, because its wide-gauge film and wide-angle lenses offered the wraparound coverage of that process without resorting to multiple cameras and projectors. (The curvature of the Cinerama screen was 146 degrees,

65 Hayes, *3-D Movies*, 51–52.

while that of Todd-AO was 128 degrees.) Described by Todd as "Cinerama out of one hole," single-strip Todd-AO required some optical manipulation in printing to avoid keystone distortion when projected on the curved screen at 2.20:1.[66] Designed for running at 30 frames per second (fps) as opposed to the standard 24 to enhance its resolution, Todd-AO proved to be a beautifully precise optical system, and its debut feature *Oklahoma!* (Fred Zinnemann, 1955)—which also featured six-track stereophonic sound—was a huge financial success. Other wide-film systems developed simultaneously with Todd-AO were Super Panavision 70, which used an unsqueezed 65mm negative for projection in either 35mm or 70mm format to produce an aspect ratio of 2.35:1 or 2.20:1, respectively (e.g., *Lawrence of Arabia* [David Lean, 1962]); and Ultra Panavision 70, which combined prismatic anamorphic and wide-gauge principles to squeeze a wide-field image onto 65mm stock. The squeeze ratio of the latter was only 1.25:1, but since the picture area was already 2.25:1, the 70mm anamorphic aspect ratio of 2.75:1 was perfect for epic spectacles, such as MGM's *Ben-Hur* (William Wyler, 1959), where it was credited as MGM Camera 65, and *Mutiny on the Bounty* (Lewis Milestone, 1962).[67] Products of the wide-film systems were typically blockbusters[68] that could be "road-shown"—toured from city to city for exclusive engagements at inflated admission prices to recoup their high production costs. Wide-film technology died out in the 1960s for the same reasons as Cinerama—it was bulky, expensive, and relatively inflexible, although it is still used occasionally today by purists like Christopher Nolan and Quentin Tarantino who seek the highest possible image quality on analog stock.

All stereoscopic features of the 1950s had 10-minute Intermission points at around the 50-minute mark or less, depending on their lengths, because

66 With this in mind, Todd had a CinemaScope version of the famous Rogers and Hammerstein musical shot simultaneously with the Todd-AO version for general release. Both are available today on DVD and Blu-ray, and comparison will show CinemaScope to be vastly inferior, both visually and aurally, to Todd-AO.

67 In 1963, Cinerama dropped its three-camera system in favor of both Ultra Panavision 70 (e.g., *It's a Mad, Mad, Mad, Mad World* [Stanley Kramer, 1963]) and Super Panavision 70 (e.g., *2001: A Space Odyssey* [Stanley Kubrick, 1968]). In the 1970s, Cinerama stopped producing films and became a distribution company.

68 The term "blockbuster" was coined during the Second World War to refer to high-capacity aerial bombs used by the RAF that could literally destroy blocks of housing at a time. It has since come to refer to big-budget, high-profile films and other cultural production designed to have broad popular appeal and make windfall profits (which

they were shown in polarized 3D on two interlocked projectors in prints representing the left- and right-eye views. At the time, the largest available projector magazines were 25 inches in diameter and could accommodate 24-inch reels which held up to 5,000 feet of color film and 5,500 of black-and-white. Furthermore, the carbon arc rods used as the projectors' light source could only burn effectively for about an hour without replacement. Most theaters maintained two or three projectors in their booths (the third, called the "dubber," was to run the three-channel magnetic soundtrack),[69] and 10 minutes was just enough time to change reels, rethread both projectors, and install new carbon rods for the picture's second half. Once screenwriters became aware of this situation, they were able to construct their scripts so that a break would occur naturally at some logical point in the plot (Figure 3.20).

The impression that 1950s 3D movies had little value beyond their exploitation of emergence effects, or negative parallax, is patently untrue. Most were carefully made in terms of their (mainly) color cinematography, layered art direction, and use of stereophonic sound, and many of them measure up to the best contemporary critical standards. The value of such high-end productions as *Kiss Me Kate* and *Dial M for Murder* is obvious, but many lower-budget genre films also have stood the test of time. *The Maze*, for example, is a model of gothic horror, and *Inferno* is a film noir whose color palette can fairly be compared to that of *Leave Her to Heaven* (John M. Stahl, 1945); *Those Redheads from Seattle* is a refreshing blend of musical and frontier western, and *Miss Sadie Thompson* is a musical comedy that bears close resemblance to *South Pacific* (Joshua Logan, 1958), in kind if not in budget; *Wings of the Hawk*, *Gun Fury*, and *Hondo* are important, if not great, westerns; *The Mad Magician* is a comedy-horror film that easily rivals its predecessor *House of Wax* and provided inspiration for many Roger Corman genre-benders; *Creature from the Black Lagoon* ventured classically luminous underwater photography and created an enduringly sympathetic monster; while *It Came from Outer Space* is one of the very best of 1950s science fiction films, exceeded only perhaps by

---

they frequently do not). The idea arose in Hollywood during the 1970s that a handful of heavily promoted blockbusters or "event movies" (now "tent-poles"), like *The Godfather*, *The Exorcist*, *Jaws*, and *Star Wars*, could subsidize a studio's failures for years to come—a calculus that has sometimes proved successful and sometimes not (see Chapter 6). Yet the logic and the practice persists, even though Netflix, Amazon Prime, and other successful streaming video services, as well as the Covid pandemic of 2020-21, have chastened it somewhat.

69 This practice caused the Projectionists' Union to demand that 3D booths be staffed by at least two of its members to handle the traffic among the three machines.

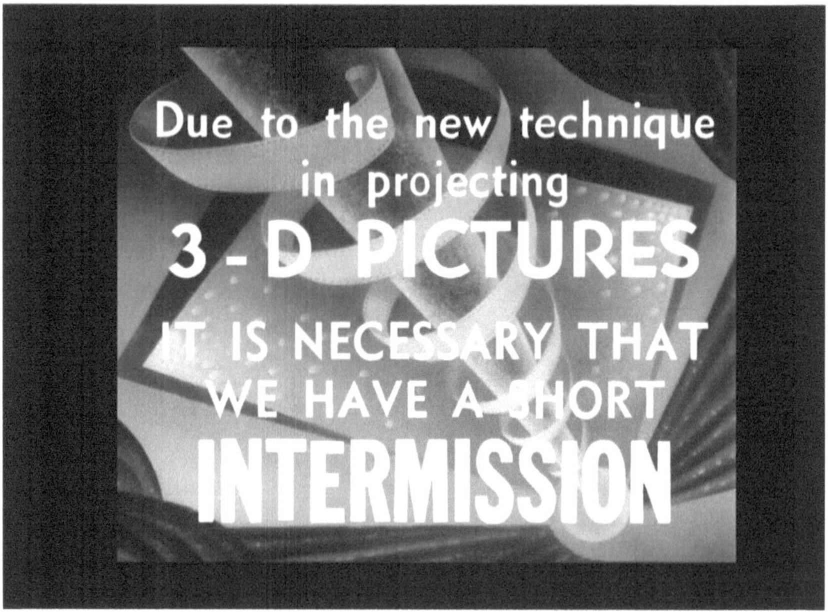

Figure 3.20 3-D Intermission Card, *c.*1953. All stereoscopic features of the 1950s had 10-minute Intermission points at around the 50-minute mark or less, depending on their lengths, because they were shown in polarized 3D on two interlocked projectors in prints representing the left- and right-eye views.

Them!, which was slated for 3D production but shot flat in 1954 because the boom was clearly over.[70]

70 Some accounts say that *Them!* was shot in WarnerVision 3-Dimension but edited flat, or perhaps finished in 3D but released flat. Whatever the case, in 1954 and forever after audiences saw the film in a black-and-white 2D version only.

Chapter 4

STEREOSCOPIC REVIVAL, 1970–85

Renewal attempts

There were several stabs at a 3D revival during the 1960s before the 1970s surge in soft- and hardcore pornography stimulated by the scrapping of the Production Code and implementation of the CARA (Code and Ratings Administration) system in October 1968. *September Storm* (Byron Haskin, 1960), independently produced by Edward Alperson and distributed by Fox, was the only film associated with the 1950s boom that was released in both Natural Vision 3-Dimension and CinemaScope, as well as in Eastman Color with prints by De Luxe, but it apparently wasn't filmed with anamorphic lenses as were contemporaneous features bearing the "CinemaScope" trademark. According to Bob Furmanek, who supervised the 3D Film Archive restoration in 2016, *September Storm* was shot full-frame using the same aspect ratio as *Bwana Devil* and *House of Wax* (1.37:1) and then extracted in postproduction to create optical prints for 2.39:1 anamorphic projection in a process similar to RKO's SuperScope (see Chapter 3). Shot on location in and around the Spanish island of Majorca, the film is a seabound adventure about a search for sunken treasure that nevertheless features impressive 3D interiors and some stunning underwater sequences; it opened in 3D in a number of major cities in September but was released flat elsewhere (Figures 4.1 and 4.2). The following year, Warner Bros. released the independently produced Canadian film *The Mask* (Julian Roffman, 1961), a fairly conventional black-and-white 2D horror thriller with wildly creative anaglyphic 3D dream sequences. The story revolves around an ancient mask, modeled on a real one used in Aztec peyote ceremonies, that causes the wearer to hallucinate murderous visions from the unconscious mind. The director had originally signed legendary Hollywood montage artist Slavko Vorkapich to design these sequences, but his script proved too expensive for the film's relatively modest budget (although he still got screen credit for it), and Roffman ended up directing them himself to create a psychedelic, occasionally disturbing viewing experience.[71] Although they were part of a William Castle-like audience participation gimmick in which spectators were given "Magic Mystic Masks" (anaglyphic 3D glasses) to see them, Roffman himself regarded the film as an avant-garde experiment, which extended to the use of electronic music composed by Myron Schaffer and Louis Applebaum to underscore its surrealistic 3D images.[72]

Historical oddities to be sure, three "adults only" 3D films appeared the following year—*Adam and Six Eves* (Roberto Rico as "John Wallis," 1962; Natural Vision 3-Dimension dual-strip 35mm), *The Bellboy and the Playgirls* (Francis Ford Coppola, 1962; dual-strip "Optivision" [16mm printed side-by-side and blown up to 35mm]), and *Paradisio* (H. Haile Chace as "Henri Haile," 1962; "Tri-Optique" dual-strip 35mm). The first concerned a prospector in the California desert who comes across six naked sunbathers, and it was released flat only in Eastman Color. The second is interesting mainly as Coppola's (and unacknowledged by him) first feature; it contains both black-and-white and color anaglyphic sequences and centers on a bellboy who spies on naked women through hotel keyholes (which is also of interest, since that can only be accomplished in fact using one eye, destroying the three-dimensional illusion). *Paradisio* was a pastiche of newly shot footage and outtakes from a 1956 Italian documentary; its main character acquires a pair of glasses that enables him to see through women's clothing (ho-hum—cf. Russ Meyer's *The Immoral Mr. Teas* [1959]) in both black-and-white and color anaglyphic stereo.

In 1966, Arch Oboler reappeared with a new single-strip over-and-under stereoscopic process, dubbed "Space Vision 4-D," with *The Bubble.* A serious science fiction feature that Obeler wrote, produced, and directed, it stands in stark contrast to the skin flick fodder of 1962. Despite major lapses in narrative logic, the film looks forward to *Westworld* as both film (1973) and cable TV series (2106–2021), as a young man and his pregnant wife traveling

---

71 This was no accident. Director Roffman, though no drug user himself, had been in contact with Dr. Timothy Leary about the effects of LSD in order to create authentic visionary images for his film. Anecdotal evidence suggests that Roffman's imagery does resemble the experience of people tripping on the substance.

72 *The Mask*, as digitally restored by the 3D Film Archive (Kino Lorber, 2015), contains 22 minutes of stereoscopic footage in three parts; a special supplement to the Blu-ray disc enables viewers with anaglyphic glasses to view these sequences as their original audiences did in 1961. There is a marginal difference in the 3D experience between the two formats—the digital restoration appears to be more "solid," while the anaglyphic is more "sparkly" and vaguely magical.

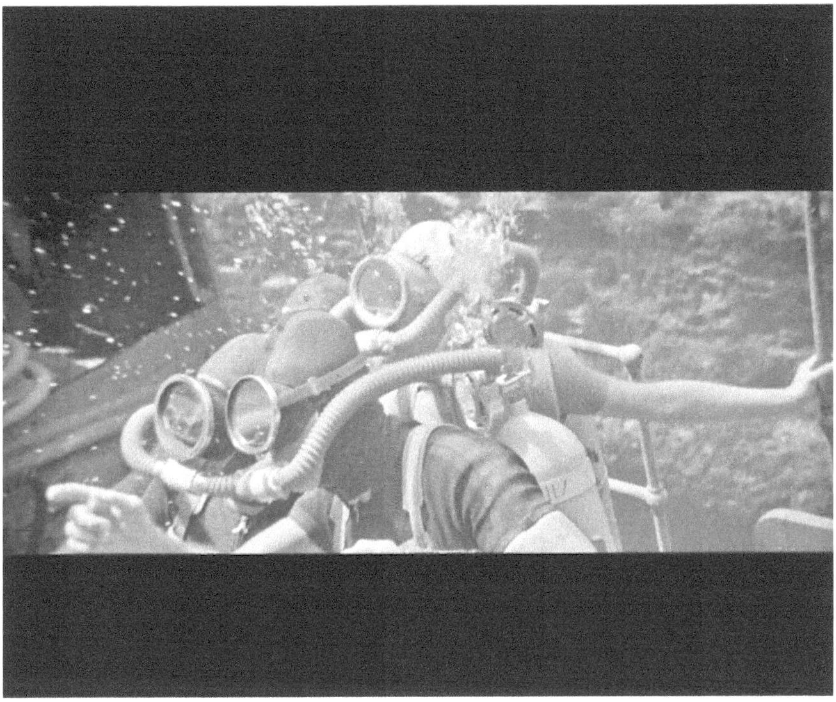

Figure 4.1 *September Storm* (20th Century Fox, 1960): shot in Natural Vision (as "StereoVision"), non-anamorphic SuperScope 235 (advertised as "CinemaScope"), Eastman Color with prints by De Luxe, and optical stereophonic sound. Hidden treasure found; underwater photography by Lamar Boren.

in a private plane are force-landed into a town peopled with (as they later say) "the walking dead"—characters on what appear to be old film sets performing repetitive tasks like robots (Figure 4.3). As silly as it sounds, the plot was very much in tune with contemporary TV fare like *The Twilight Zone* (1950–64) and *One Step Beyond* (1959–61), and the protagonists eventually realize that they have been placed in a "human zoo" by aliens who have trapped the whole community in an impenetrable plastic dome for observation. *The Bubble*, photographed in Eastman Color by Charles F. Wheeler, contains artful art direction (by Marvin Chomsky, also associate producer) and many impressive emergence effects, including a variety of overhead shots in which objects pop out of the screen at an 180-degree angle to the viewing plane. It also contains some surprisingly prescient observations on alien intelligence that anticipate much more sophisticated science fiction films like *Arrival* (Denis Villeneuve, 2016): at one point, the husband remarks of the anthropomorphic fallacy, "Even God was supposed to look like us" and, as to the suddenness with which

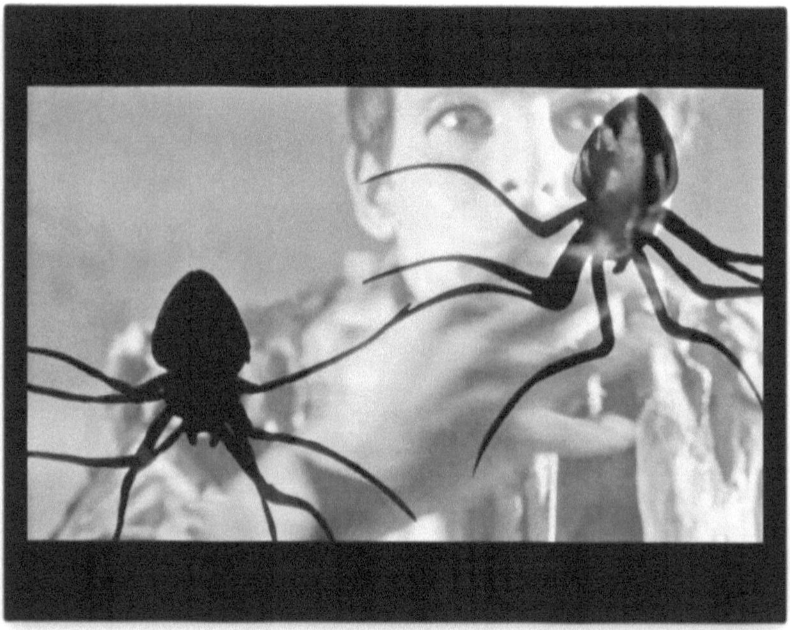

Figure 4.2 *The Mask* (Warner Bros., 1961): shot in "Depth Dimension" (dual 35mm single-strip anaglyphic) and presented in black-and-white Widescreen (1.85:1), with optical "Electro-Magic Sound," based on electronic feedback. From one of the four hallucinatory sequences that could only be seen by donning anaglyphic 3D "Magic Mystic Glasses."

the bubble has entrapped them, "Why should time be the same to them as it is to us?" (The script also pushed the PCA envelope with frank references to contraception, breastfeeding, prostitution, and the use of the term "Bastards!" to describe the aliens.) Like *September Storm*, *The Bubble* was released in the CinemaScope aspect ratio of 2.35:1 but through a different process: the Space Vision camera's over-and-under lens configuration cut the standard 35mm frame in half to achieve a Scope ratio without anamorphic distortion, similar to Techniscope;[73] it originally ran 112 minutes, but was edited down to 93 and

73 Introduced by Technicolor Italia in 1960, Techniscope used a two-film perforation negative pulldown per frame, instead of the standard four-perforation frame exposed in conventional 35mm cinematography. It was an inexpensive alternative to other widescreen processes because it used half the amount of film stock and a non-anamorphic spherical lens. Release prints were made by anamorphizing and

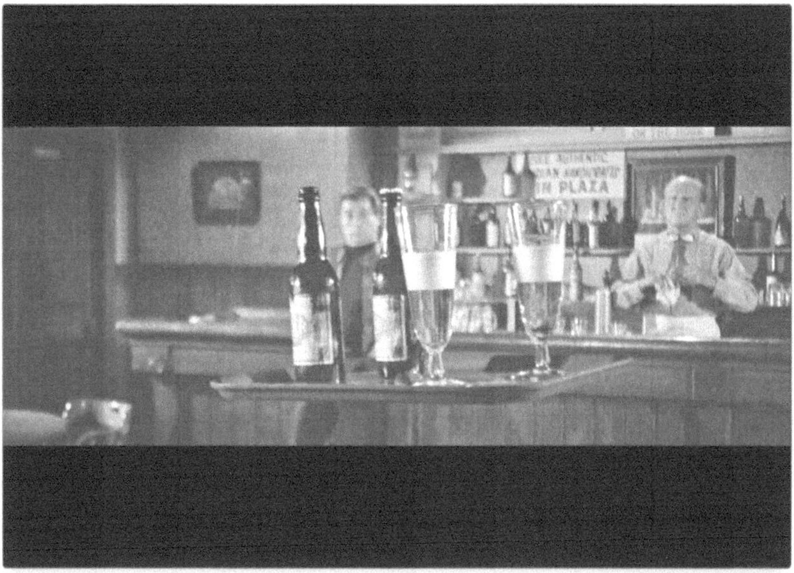

Figure 4.3 *The Bubble* (Arch Oboler Productions, 1966): shot in "Space-Vision 3-D," a new single-strip over-and-under 2.35:1 process developed by Natural Vision's Arch Oboler that became the basis for shooting and exhibiting stereoscopic films for the next 30 years. Here, a tray of beers floats before an amazed civilian pilot and a zombified bartender.

rereleased in 1976 as *Fantastic Invasion of Planet Earth*. In neither version did the film amortize its reported budget of $500,000.

Sex and gore in 3D

None of the 1960s movies mentioned above, however, had any significant impact on the future of three-dimensional cinema. It was the 1969's *The Stewardesses* that did the trick. Directed by Alf Silliman Jr. (Allan Silliphant, who also wrote the screenplay), shot in StereoVision (single-strip side-by-side

doubling the size of each frame, which gave the format 50 percent of the resolution of full-frame 35mm and a sort of gritty, documentary look. More than 350 films were shot in Techniscope between 1960 and 1980, most of them European productions in the horror and western genres (e.g., *gialli* and spaghetti Westerns), although some Hollywood studios used it for low-budget projects and Universal used it extensively from the mid-1960s to the early 1970s (e.g., George Lucas's *THX 1138* [1971] and *American Graffiti* [1973] were both shot in Techniscope).

anamorphic 1.33:1, originally credited as Magnavision 3-D) and Eastman Color, and blown up to 70mm for distribution, the film is basically just a feature-length anthology of couples copulating or stimulating themselves sexually (together or in private) in variously explicit, but non-hardcore (no visible penetration) ways. Running 93 minutes in its restored version (up from 75 minutes in 1969 and 83 in 1972),[74] *The Stewardesses* is a mindless, boring film, but it is the first 3D production to offer naked female bodies, including pubic hair, in negative parallax, fulfilling the coy décolleté promise of so many "serious" 1950s boom films (as in, "this-is-what-you-wanted-in-the-first-place-and-now-you've-got-it"), and in some markets it became one of the top-grossing films of the year (Figure 4.4). In fact, *The Stewardesses* had a fabled two-year run from 1970 to 1972, returning $26 million on its $100,000 investment, and it led to a slew of sequels and imitations. Historically, it was the highest earning 3D film before *Avatar* (James Cameron, 2009) and the first American 3D film to be shown in 70mm, however spotty the presentation. In terms of return on investment, in fact, *The Stewardesses* actually far exceeded that of *Avatar*, even though the latter became for a decade the high-grossing film in the history of cinema.[75] Cameron's film, produced for $425,000,000, returned $2,789,679,794 worldwide, a ratio of about 6.5:1. The ratio for *The Stewardesses* was 260:1, convincing many potential investors that the future of 3D—indeed, that the future of movies generally—lay in pornography.

Another 3D feature of 1969 pointed in a different but equally lurid direction. This was the Spanish-German coproduction *La marca del hombre lobo* (*The Mark of the Wolfman* [Henry L. Egan (Enrique L. Equiluz)]), scripted by and starring Jacinta Molina Alvarez, aka Paul Naschy, as the tragic Count Waldemar Daninski in the first of a series of eight blood-splattered werewolf movies made in and around Madrid from 1969 through 1983. Badly cut and released in the United States as *Frankenstein's Bloody Terror*[76] in 1972, it was

---

74 The 3-D Film Archive presented a restored 93-minute version of *The Stewardesses* in a 2009 Kino Lorber Blu-ray. Intriguingly, it was paired with the considerably more explicit (though still not hardcore) 27-minute Eastman Color short *Experiment in Love* (aka *The Starlets*, 1977; "Quadravision 4-D" [Stereovision side-by-side anamorphic 1.33:1], which uses the conceit of two nearly naked women trying to calibrate a 3D camera rig for parallax convergence as part of an academic project; ultimately a handsome young gardener helps them consummate the project).

75 *Avatar* was edged out by *Avengers: Endgame* (Anthony and Joe Russo) in 2019, another 3D blockbuster, whose worldwide gross was $2,797,800,564.

76 There is a brief reference to the Frankenstein family in the prologue claiming it was cursed by lycanthropy, after which the name is never heard in the film again.

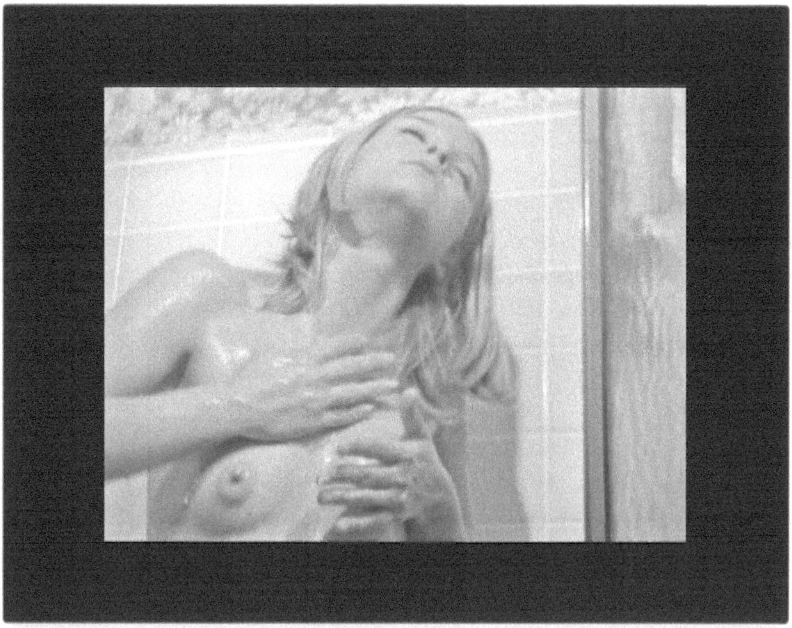

Figure 4.4 *The Stewardesses* (Sherpix Inc., 1969): shot in StereoVision (single-strip side-by-side anamorphic 1.33:1, originally credited as Magnavision 3-D) and Eastman Color, with sound by RCA Photophone. Three-dimensional cinema fulfills its erotic promise.

shot in an over-and-under 3D process in anamorphic 35mm and blown up to 70mm for distribution with the tag line "Filmed in Super 70mm Chill-O-Rama." *The Stewardesses* and *La marca del hombre lobo* set the mold for exploitative 3D in the 1970s. In 1971, for example, two films appeared with anaglyphic 3D sequences cut into preexisting 2D titles: *Asylum for the Insane* (Donald E. Davidson) was in fact the 1967 feature *She Freak* (Byron Mabe), a fumbling remake of Tod Browning's 1932 classic *Freaks*, with new footage added in duo-color anaglyphic form, while *Secrets of Ecstasy '72* was a largely flat, black-and-white porn film (in popular terms, a "skin flick") made with the same 16mm rig as *The Bellboy and the Playgirls*, but without the services of Francis Ford Coppola.

Ray Zone has noted that *The Stewardesses* and its ilk capitalized on a unique historical moment shaped by the countercultural revolution of the late 1960s and the liberality of the new CARA rating system, which by segregating audiences demographically paved the way for more adult content in American films.[77] In the porn category fell a number of adult 3D features produced by Steven Gibson and Arnold Herr in a process branded as Deep Vision. This was a polychromatic (full-color) anaglyphic format that composited the left- and right-eye views through a beam-splitter patented by Gibson onto a single strip of Eastman Color negative stock; it used a single projector and did not require special lenses or a silvered screen. For just over a decade, Gibson and Herr produced Deep Vision films with titles like *The Playmates* (Pierre La Farce [Steven Gibson], 1973), *Black Lolita* (Steven Gibson, 1975; rereleased as *Wildcat Woman* with new footage in 1975), *The Lollipop Girls in Hard Candy* (Norm De Plume [Steven Gibson], 1976), *The Disco Dolls in Hot Skin* (uncredited, 1978), *Scoring!* (Steven Gibson, 1984), and *Blonde Emmanuelle* (Giorgio Ferrari, 1984). Many of these were released in different versions for different markets—R-rated (mature adults), X-rated (softcore), and self-imposed XXX (hardcore). It is well to remember that this was the era of "porno chic" in 2D as well as 3D, when movies like *Deep Throat* (Gerard Damiano, 1972), *Behind the Green Door* (Artie and Jim Mitchell, 1971), and *The Devil in Miss Jones* (Gerard Damiano, 1973) became major box-office hits. As Jon Lewis points out, these three hardcore productions alone outearned big-budget studio films on both a screen-by-screen basis and total box-office revenue.[78] In just over four years from the institution of the CARA system, the graphic depiction of sex had become one of the most popular attractions in American cinema, and XXX-rated features accounted for 3 of the 15 most profitable films of the 1972–73 season.[79]

Some other 3D adult erotic titles of the period were *Swingtail* (Dave Shane, 1969; shot in "Cosmovision"—single-strip duo-color anaglyphic), *Prison Girls* (Thomas De Simone, 1972; shot in Optovision—dual 16mm over-and-under, printed up to single-strip 35mm), the British-made *Three Dimensions of Greta* (Pete Walker, 1973; 3D sequences in Spacemaster—dual 35mm printed single-strip anaglyphic), the West German feature *Love in 3-D* (Walter Boos, 1974; filmed in Triarama—70mm single-strip anamorphic side-by-side, optically converted to over-and-under 35mm 2:1), *International Stewardesses* (Alf Silliman Jr. [Allan Silliphant 1974]; a sequel to *The Stewardesses* shot in the same StereoVision process),[80] *Funk in 3D* (Michael Findlay, 1976; filmed in Super-Touch 3-D, the

77 Zone, *3D Revolution*, 104.

78 Jon Lewis, "Money Matters: Hollywood in the Corporate Era," in *The New American Cinema*, Jon Lewis, ed. (Durham, NC: Duke University Press, 1998), 91. See also Jon Lewis, *Hollywood V. Hardcore: How the Struggle over Censorship Created the Modern Film Industry* (New York: New York University Press, 2002).

79 David A. Cook, *Lost Illusions: American Cinema in the Shadow of Watergate and Vietnam* (New York: Scribner's, 2000), 276.

80 Aka *Supersonic Supergirls*, this film was reedited by the producers and rereleased with new footage variously as *SST* or *The Stewardesses, Part II* in 1984, and as *Airline* or *The*

single-strip over-and-under 2.35:1 system later used for the 1981 *Comin's at Ya!*), Sweden's *A Man with a Maid* (aka *The Groove Room*, Vernon P. Becker, 1977; shot in a single-strip side-by-side vertical process called "Space Cinema Camera 3D Wondavision"),[81] *The Capitol Hill Girls* (William J. Condos [Bill Condon], 1977; filmed in "Polarized LazerVision"—another name for StereoVision side-by-side anamorphic 1.33:1),[82] *The Surfer Girls* (Frank Sillman [Allan Silliphant], 1978; filmed in StereoVision), and—into the 1980s—the French-produced *Venus* (Peter Hollison, 1984; shot in StereoVision anamorphic 2.35:1) and *Emmanuel IV* (Francis Leroi and Iris Letaans, 1984; filmed in Arrivision [Deep Vision rebranded] and StereoVision 1.66:1). There were in addition several hardcore features that targeted gay males—such as *Ram Rod* (uncredited, 1973; filmed in single-strip side-by-side anamorphic 35mm), *Manhole* (uncredited, 1977; shot in Marks 3-Depix single-strip over-and-under), and *Heavy Equipment* (Lancer Brooks, 1977; filmed in 3-Dimensions single-strip anaglyphic). For most of the 1970s, however, the primary audience for adult films—in 3D and otherwise—was heterosexual couples, who indulged in "porno chic" entertainment due to its novelty, as a source of erotic stimulation, and/or out of simple curiosity. Writing in *Esquire*, Nora Ephron said of the moment, "not to have seen it [*Deep Throat*] seemed somehow derelict."[83]

Conceived in the same grisly horror mode as *La marca del hombre lobo* was Andy Warhol's X-rated *Flesh for Frankenstein* (Paul Morrisey and Anthony M. Dawson [Antonio Margheriti], 1973)—in which internal organs and body parts assaulted the audience in Spacevision 3D (single-strip over-and-under 2.35:1), combined with much nudity and frenzied sexual activity. Aka *Andy Warhol's Frankenstein*, the film was produced by Carlo Ponti, shot in Italy with special effects by future Oscar-winner Carlo Rambaldi,[84] and it contains many remarkable emergence effects. In fact, as R. M. Hayes notes, "the entire film was shot as if all the action were appearing in the theater space."[85] Despite its relatively low budget, *Flesh for Frankenstein* had good production values,

---

Stewardesses, Part III in 1987. This kind of cannibalization was (and still is) common among exploitation producer/distributors.

81 A loose, softcore adaptation of a famous English-language pornographic novel, this film was reedited and rereleased variously as *What the Swedish Butler Saw* (1981), *Teenage Tickle Girls* (1983), and *Tickled Pink* (1983).

82 Rereleased variously as *The Arab Connection* (1978), *The Best Little House on Capitol Hill* (1981), *Capitol Hill* (1984), and *Behind Closed Doors* (1989).

83 Quoted in Linda Williams, *Hard Core*, 99.

84 Rambaldi won a Special Achievement Oscar for Dino De Laurentiis's 1976 remake of *King Kong* and Academy Awards for Visual Effects as part of creative teams for *Aliens* (Ridley Scott, 1979) and *E.T.: The Extraterrestrial* (Steven Spielberg, 1981).

85 Hayes, *3-D Movies*, 187.

received largely positive reviews, and was the second most commercially successful 3D film to date after *The Stewardesses*.[86] Not so England's *The Flesh and Blood Show* (Pete Walker, 1974; anaglyphic duo-color cropped to 1.75:1) and South Korea's *Ape* (aka *A*P*E*, Paul Leder, 1976; filmed in Spacevision single-strip over-and-under 2.35:1). The former was a lurid 2D gore-fest set in a seaside theater with a 14-minute 3D conclusion, and the latter a cheap $25,000 *Godzilla*-style monster move attempting to cash-in on the upcoming release of Dino De Laurentiis's big-budget remake of the 1933 *King Kong* (John Guillerman, 1976).[87]

Generic horror and science fiction in 3D

Like pornography, horror and science fiction were seen as exploitation genres before the 1970s, but all three were eminently suitable to 3D. (Whereas porn had never been respectable, horror and science fiction—different genres, to be sure—had distinguished roots in German Expressionism but became associated with low-budget sensationalism after the Second World War.) A number of important horror films appeared in 3D following the box-office success of Ferdinando Baldi's stereoscopic western *Comin' at Ya!* in 1981. *Parasite* (Charles Band, 1982), for example, filmed in StereoVision (single-strip over-and-under 2.35:1), had a solid script, creative emergence effects, and an SFX team headed by Stan Winston; it concerns a scientist in a postnuclear America *c.*1992 who breeds leach-like parasites that grow every time they feed, one of which attaches itself to his stomach (Figure 4.5). At the same time, a number of popular horror franchises adopted the 3D format for the third entry in their ongoing series: *Friday the Thirteenth—Part 3: 3D* (Steve Miner, 1982),[88] *Jaws 3-D* (Joe Alves, 1983), and *Amityville 3-D* (Richard O. Fleischer, 1983) (Figures 4.6–4.8). All three were shot in some version of Arrivision 3-D, a single-strip over-and-under 2.39:1 system that could be used with anamorphic lenses; distributed by major studios—Paramount, Universal, and Orion, respectively—they were the first wide releases of single-strip 3D features, and all were commercially successful to a degree relative to their

---

86 *Frankenstein* was shot back-to-back in Italy with Andy Warhol's *Blood for Dracula* (Paul Morrisey, 1974), featuring nearly the same cast and crew. Originally planned for 3D, Dracula was shot and released flat in spherical 1.85:1.

87 The producers of *A*P*E* added asterisks to the title in aspirational imitation of the box-office hit *M*A*S*H* (Robert Aldrich, 1970), since much of the footage featured training maneuvers by the South Korean army. Paul Leder (1926–1996) was a legendary exploitation writer/producer/director whose *I Dismember Mama* (1972) and *The Baby Doll Murders* (1993) are classics of the form.

88 Also spelled *Friday the 13th* onscreen and in some promotional materials.

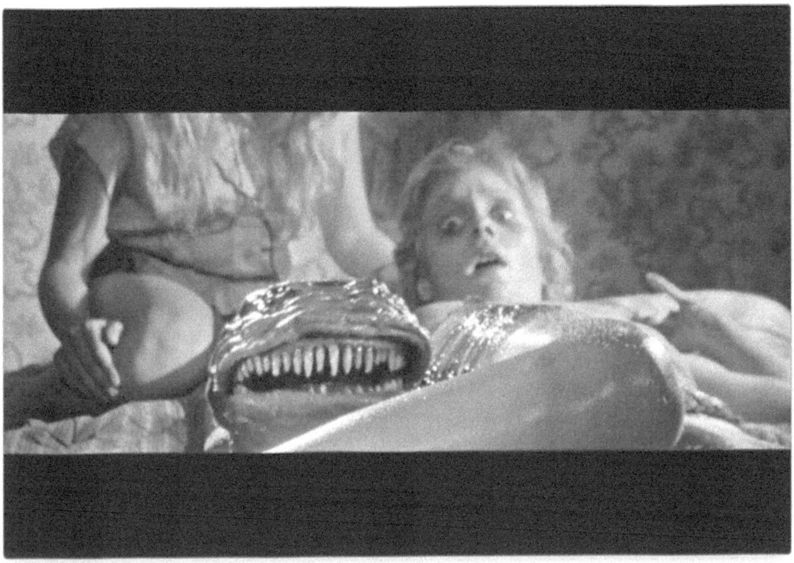

Figure 4.5 *Parasite* (Embassy Pictures, 1982): shot in StereoVision (a version of Space-Vision, single-strip over-and-under 2.35:1), with Eastman Color ("Metrocolor") and sound by RCA Photophone. A parasite is born: special effects by Stan Winston.

budgets.[89] (The most so was *Friday the Thirteenth—Part 3*, returning $36,690,000 against a $2,300,000 investment; next was *Jaw 3-D*, produced for $20,500,000 and grossing $45,518,000; *Amityville 3D* basically broke even, earning $6,353,135 against its $6,000,000 budget.) Lower on the scale were Lucio Fulci's *Zombie 3-D* (1984; Arrivision), the anthology film *Tales of the Third Dimension* (Todd Durham, Worth Keeter, and Thom McIntyre, 1984; StereoVision), and slasher *Silent Madness* (Simon Nutchtern, 1984) (Figure 4.9).

Thanks to the phenomenal success of George Lucas's *Star Wars* in 1977, popular science fiction—which during the 1950s had become associated with

89 All three were viewed through polarized glasses in theatrical distribution; so are the Blu-ray releases of *Jaws 3-D* (Universal, 2016) and *Amityville 3-D* (MGM/Scream Factory, 2013), but that of *Friday the 13th Part 3: 3-D* (Paramount) is anaglyphic. *Jaws 3-D* was released with a View-Master program as a promotional tie-in. Additionally, a battery-powered stereoscopic video game based on the film was released by Tomy in Japan as *Jaws 3D Graphic Game* and in the United States as *Tomytronic 3D: Shark Attack*. See Leon Gurevitch and Mirian Ross, "Stereoscopic Media: Scholarship Beyond Booms and Busts" in *3D Cinema and Beyond*, Public 47 (Toronto: Ontario Arts Council, 2013), 83–93.

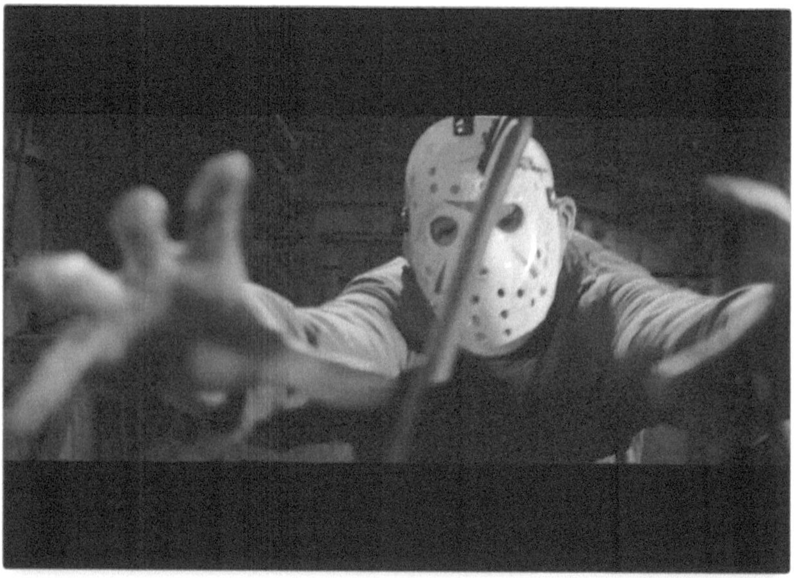

Figure 4.6 *Friday the Thirteenth*—Part 3: 3D (Paramount Pictures, 1982): shot in 3-Depix (another version of Space-Vision, single-strip over-and-under 2.35:1), with Eastman Color ("Movielab") and optical sound. Jason gets the axe and reaches into audience space.

B-grade alien-invasion and monster films[90]—became viable as mass entertainment again. With its emphasis on space flight and spectacular special effects, the genre was naturally suitable to 3D. The first 1980s three-dimensional science fiction film was Columbia's *Spacehunter: Adventures in the Forbidden Zone* (Lamont Johnson, 1983), shot and released in StereoVision 2.35:1 (also released flat at 1.85:1). Produced by Ivan Reitman (*Ghostbusters*, 1984) and scored by Oscar-winner Elmer Bernstein,[91] the film concerns a Han Solo-like bounty hunter on a mission to rescue three female earthlings from an evil warlord on a planet inhabited by mutants; it was not a hit in contemporary

90 There were exceptions, of course—for example, *The Thing* (Christian Nyby, 1951), *The Day the Earth Stood Still* (Robert Wise, 1951), *When Worlds Collide* (Rudolf Mate, 1951), *War of the Worlds* (Byron Haskins, 1953), *Them!* (Gordon Douglas, 1954), and *Forbidden Planet* (Fred Wilcox, 1956). Stanley Kubrick's *2001: A Space Odyssey* (1968) also helped to redeem science fiction, but, like most of the director's work, it was seen to be sui generis rather than generic.

91 Bernstein won an Academy Award for Best Original Music Score for *Thoroughly Modern Millie* (George Roy Hill, 1967); he also won Golden Globes for the scores of *To Kill a Mockingbird* (Robert Mulligan, 1962) and *Hawaii* (George Roy Hill, 1966).

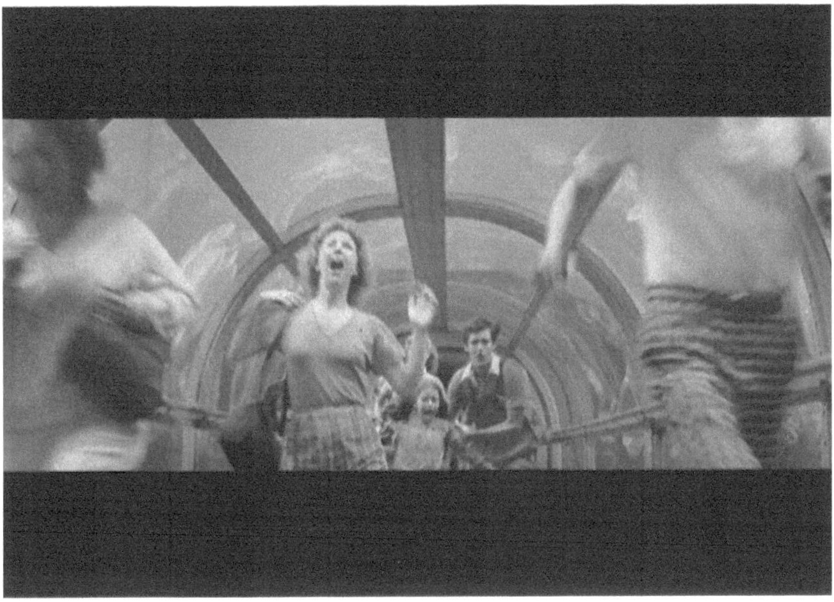

Figure 4.7 *Jaws 3-D* (Universal Pictures, 1983): Shot in Arrivision (yet another version of SpaceVision, single-strip over-and-under 2.35:1; reframed to 2.50:1), Eastman Color, and Dolby Stereophonic Sound. Waterpark crowd runs in panic to escape imploding undersea tunnel.

terms, earning just $16.5 million against a $14.4 million budget. Shortly thereafter, Universal released *Metalstorm: The Destruction of Jared-Syn* (Charles Band, 1983), which identified itself in advertising as a "space-age western" and was very similar to *Spacehunter* in its tale of a "warrior ranger" in pursuit of an intergalactic criminal who leads a race of mutants on a desert planet (Bronson Canyon in Griffith Park, Los Angeles) evocative of the then-popular *Mad Max II: Road Warrior* (George Miller, 1981) landscape (Figure 4.10). Shot in StereoVision 2.35:1, complete with a magic mask and a hidden city, the film was less well produced but ultimately more popular than its processor.[92] Even less distinguished was *Hyperspace* (aka *Gremloids*, Todd Durham, 1984), an Earl Owensby Studios (EO)[93] feature, shot in StereoVision 2.35:1, which was

92 *Metalstorm* is exactly 87 minutes long and has one of the longest end-credit sequences in pre-CGI history—a full eight minutes on the Shout Factory 2016 Blu-ray edition, which is particularly notable given its low-end makeup and special effects.

93 Founded in 1974 in Shelby, North Carolina, Earl Owensby Studios (EO) produced dozens of low-budget action programmers for distribution to drive-ins, grindhouses, and nascent cable TV during the 1970s and 1980s.

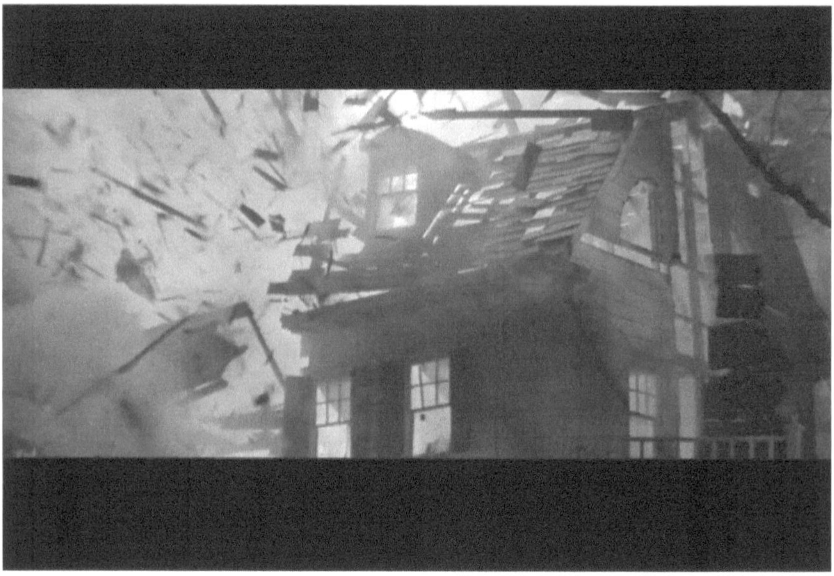

Figure 4.8 *Amityville 3-D* (Orion Pictures, 1953): shot in ArriVision (single-strip over-and-under 2.35:1), Eastman Color with prints by De Luxe, and Lion's Gate stereophonic sound. The family home explodes at the conclusion, flinging stereoscopic debris at the audience.

a low-end science fiction comedy featuring a bumbling alien warlord who has a series of misadventures after he accidentally lands on earth. The last stereoscopic science fiction film of the 1980s, *Starchaser; The Legend of Orin* (Stephen Hahn, 1985) was also promoted as the first animated 3D feature, although in fact the Australian *Abra Cadabra* (Alexander Stitt, 1983) had preceded it by several years. A South Korean-US coproduction distributed by Atlantic Releasing, *The New York Times* called this film "such a brazen rip-off of George Lucas's *Star Wars* that you might think lawyers had been called in."[94] The same might be said of all of the 3D, as well as most 2D, science fiction films of this period, so great was the influence of *Star Wars* on the genre.[95]

94 Vincent Canby, "Screen: 'Starchaser,' An Animated Space Tale," *New York Times*, 1985.

95 Exceptions would be Ridley Scott's *Alien* (1979) and its sequel James Cameron's (*Aliens*, 1984).

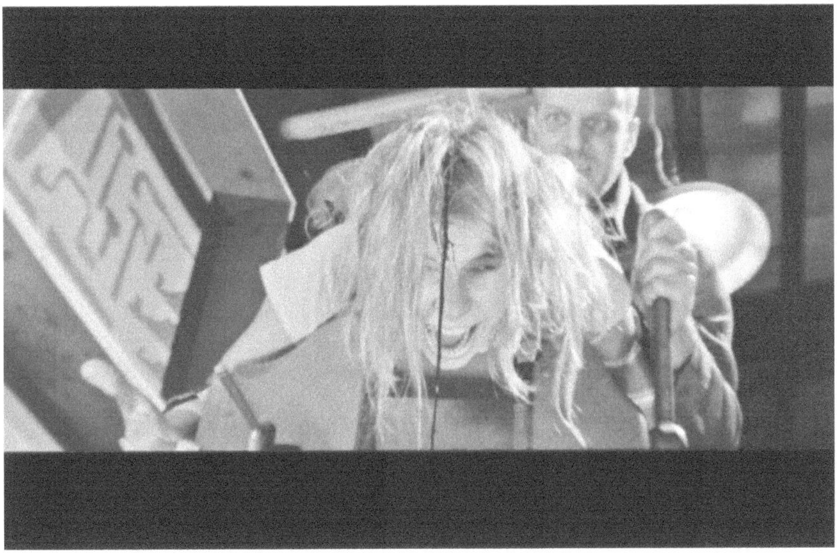

Figure 4.9 *Silent Madness* (Almi Pictures, 1984): shot in ArriVision, Eastman Color, and Dolby Stereophonic Sound. The madman (Solly Marks) crushes a victim's skull in negative parallax.

Other exploitation genres

It was action-adventure that actually catalyzed 3D filmmaking in the 1980s, starting with the Spanish-Italian-America coproduction *Comin' at Ya!* (Ferdinando Baldi, 1981), filmed in Eastman Color, widescreen, and stereophonic sound. The stereoscopic process, an over-and-under single-strip 35mm format, was unique: two Techniscope frames, one for the left-eye image and one for the right, were stacked one above the other in the same area as one ordinary CinemaScope frame yielding an aspect ratio of 2.35:1.[96] The lens system used was Bill Bukowski's Optimax III, known for introducing vertical parallax due to its flawed design that played hell with convergence. Although this was a problem for the picture in general, it was corrected in theaters by placing "mirror box" prismatic converters in front of an ordinary spherical projection lens which were meant to converge the stacked images and cross-polarize them to match the filters in the 3D glasses.[97]

---

96 A prototype of the system had been used once before, as "Super-Touch 3D," in Michael Findlay's *Funk in 3D* (1976)—see above.

97 Convergence and depth of field were still issues in the 2016 Blu-ray restoration of *Comin' at Ya!* by MVD Visual, although its stereo presentation was strong.

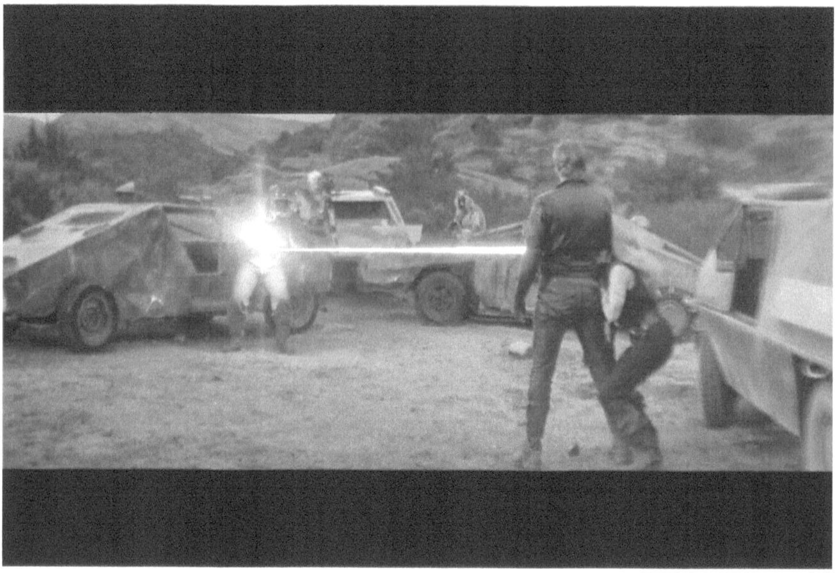

Figure 4.10 *Metalstorm: The Destruction of Jared-Syn* (Universal Pictures, 1983): shot in StereoVision (single-strip over-and-under 2.35:1; reframed to 2.50.1), Eastman Color with prints by Technicolor, and Dolby Stereophonic Sound. Ranger Finder Dogen (Jeffrey Byron) blasts a minion of Jared-Syn.

Distributed by Filmways, the successor company to AIP, the film itself was a violent if good-spirited spaghetti western (and, in fact, was a remake of Baldi's own 1971 *Blindman*),[98] full of Peckinpah-style death balletics, and it became quite popular, returning $12 million on a modest $2.5 million budget (Figure 4.11). Two years later the same production team made *Treasure of the Four Crowns* (Fernando Baldi, 1983), a sprawling adventure film that mimicked Steven Spielberg's *Raiders of the Lost Ark* (1981) to the point of borrowing some of its key action sequences. Like *Comin' at Ya!*, the 3D process was billed as "SuperVision" and "WonderVision," but *Four Crowns* was in fact shot using the Marks 3-Depix Stereospace Converter, a version of the Arrivision camera system employed for *Friday the 13th, Part 3* that produced an aspect ratio of about 2.4:1. Despite its many exciting emergence effects and dynamic Dolby surround sound, the movie was so derivative that it garnered negative reviews and did poorly at the box office.

98 Baldi (1917–2007) was a veteran Italian exploitation director, having worked in peplum (*David and Goliath* [1960]), spaghetti westerns (*Texas, Adios* [1966]; *Django, Prepare a Coffin* [1968]), and crime films (*The Sicilian Connection* [1972]).

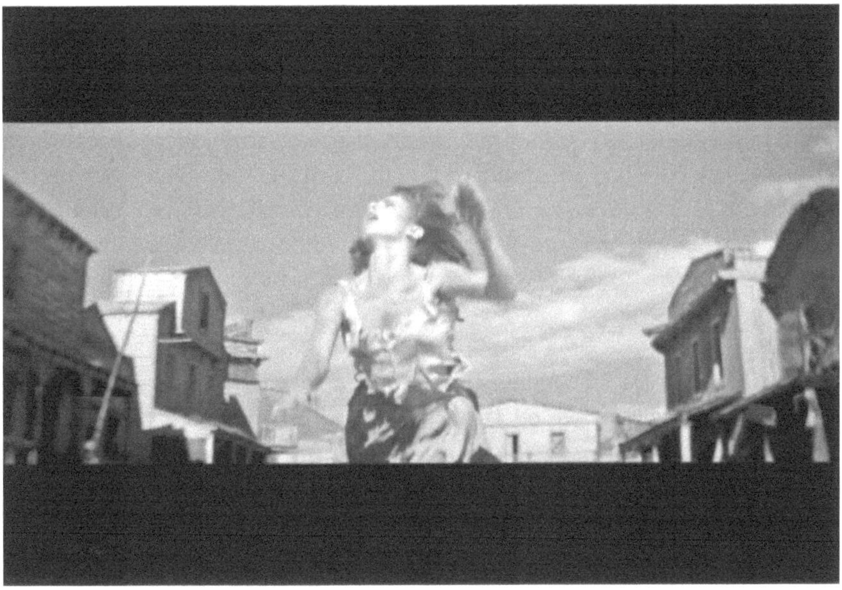

Figure 4.11 *Comin' at Ya!* (Filmways Pictures, 1981): shot in Optimax III (single-strip over-and-under 2.35:1) and presented in Dimensionscope 3-D (a version of Techniscope), Eastman Color with prints by Technicolor, and Dolby Stereophonic Sound. Victoria Abril is shot in the back and will fall into negative parallax.

The other stereoscopic action films of the 1980s were shoestring budget affairs, shot in StereoVision and produced by Earl Owensby Studios (EO). The best of these was *Rottweiler* (aka *Dogs of Hell*, Worth Keeter, 1982), where a pack of rottweilers trained by the US military to kill humans, escapes to terrorize the mountain resort town of Lake Lure, North Carolina. Despite its obvious lack of resources, the film had top-notch 3D effects assisted by StereoVision consultant Chris Condon and is regarded as one of the best such films of its era. EO's other 3D action films were *Hit the Road Running* (Worth Keeter, 1983), *Chain Gang* (Worth Keeter, 1984), and *Hot Heir* (aka *The Great Balloon Chase*, Worth Keeter, 1984); all had limited distribution and little impact beyond the southeastern drive-in circuit.

"Kung fu" (literally "achievement" or "merit" in both Cantonese and Mandarin) was another 1970s exploitation genre that lent itself readily to stereoscopic presentation. At the time, martial arts films from Hong Kong, flippantly known as "chop-socky," appealed to a broad sector of nonwhite moviegoers—including Asian Americans but also Hispanics and, especially, inner-city blacks—perhaps because the central theme of many of them was defense of one's own ethnic group against another. The primary appeal of kung fu films, however, was their skillfully shot and edited fight sequence, which were performed and choreographed by real Chinese martial arts masters according to centuries-old disciplines of self-defense. The first 3D marital arts film was *Revenge of the Shogun Women* (aka, *Thirteen Golden Nuns*, Mai Chun Chang, 1977), shot in a single-strip over-and-under process using the Optimax III system to produce a 2.39:1 aspect ratio. Rife with nudity and violent action, *Revenge* concerns a group of women who have been gang-raped by bandits in nineteenth-century China and who subsequently train to become nuns. *Dynasty* (Mai Chun Chang, 1977), from the same Hong Kong/ Taiwanese production company (Eastern Media), was shot in a similar 3D process with eight-track quadraphonic sound (i.e., Sensurround). In it, a royal prince must fight for life against accusations of treason, and it contains a riot of in-your-face emergence effects as swords, arrows, spears, and body parts hurtle from the screen into audience space. Taiwan's *The North and South Chivalry* (Yung-Hua Liu, 1978) and Hong Kong's *The Magnificent Bodyguards* (Lo Wei, 1978) were both shot in dual Techniscope single-strip over-and-under. The first had only limited distribution outside of Taiwan, but *Bodyguards* became a hit in Hong Kong, primarily because it starred Jackie Chan, who also worked as the fight coordinator.

The only other notable 3D feature released in the United States during the 1980s was Paramount's one-off spy comedy *The Man Who Wasn't There* (Bruce Malmuth, 1983),[99] shot with the Optimax III single-strip system, about a secret agent who discovers a formula for invisibility and becomes the target of international espionage; it is notable mainly for some gratuitous R-rated nudity and an early leading performance by Steve Guttenberg. In related activities, several classic stereoscopic films from the past were rereleased theatrically during the period, most of them in single-strip anaglyphic duo-color form: in 1971, *House of Wax* (1953); in 1972, *It Came from Outer Space* (1953) and *Creature from the Black Lagoon* (1954); in 1977, *The Bubble* (reedited and retitled as *Fantastic Invasion of Planet Earth*, 1977); and in 1983, *Dial M for Murder* (1954). There were also several experiments in televising 3D films processed by 3D Video Corporation during the decade. For example, *Gorilla at Large* was syndicated in several urban markets (e.g., Atlanta) in the summer of 1982 as part of a special promotion with Revco pharmacies and 7/11 convenience stores, which provided viewers with free cardboard anaglyphic glasses that failed to synchronize with most receivers (and were only marginally successful at converging 3D imagery when they did).

Driven initially by softcore and then hardcore porn, stereoscopy in the 1970s and early 1980s produced a good deal of exploitative trash, but also

99 Not to be confused with the Coen Brothers' (2001) film noir of the same title.

some notable creative hits, especially in the area of existing genre franchises. In general, 3D was an inexpensive, though not necessarily easy, way to enhance medium- to low-budget generic entertainment. Even though they had problems of convergence caused by lens distortion and fixed interaxial distances (which, e.g., made miniatures difficult to blend with live action), the most important thing about these single-strip over-and-under productions was that they kept three-dimensional cinema alive until it could reemerge in the mid-1980s in a more serious, technologically sophisticated, and culturally respectable form.

A chorus line

It will be understood by now that stereoscopy, going back its 19th photographic incarnations, has consistently fetishized women's bodies. During the 1950s boom, it is hard to find a movie in any genre that does not feature objectified, if costumed (sometimes discreetly, sometimes not so much), feminine breasts, buttocks, and—because of the rigors of the PCA—legs in chorus lines or similarly revealing presentations.[100] And as the film industry moved toward the end of the Code, there was a prevalence of "crotch shots" in bathing suits (*September Storm*) and peek-a-boo dance-hall girl costuming (*The Bubble*), so it was almost inevitable that when the Code was replaced by the Ratings System in 1968, the potential for 3D to create male-oriented pornography—both heterosexual and gay—would be at least partially fulfilled. One could argue that three-dimensional representation has always had a dark, pornographic underside, and that would be true. But, then, so too have all forms of representation since classical antiquity (when, sometimes, as in Roman phallic statuary, it wasn't dark at all but simply normative).

The fact that 3D cinema was sustained through the 1970s and 1980s mainly by sexual exploitation shouldn't dissuade us from its importance. If it weren't for *The Stewardesses* we probably would not have *Avatar* (maybe not such a bad thing) but then neither would we have *Hugo*, *Pina*, *The Hobbit* trilogy or *They Shall Not Grow Old*, which would create a terrible void in film history. In criticism of the arts, moral judgement is not simply bad practice—it is fatal.

100 I am hardly the first film scholar to notice this. See, for example, Ariel Rodgers, *Cinematic Appeals: The Experience of New Movie Technologies*, Chapter 6, "Points of Convergence," pp. 179–225 (New York: Columbia University Press, 2013).

Chapter 5

THE AGE OF IMAX, OR THE "IMMERSIVE CINEMA," 1986–2009

The Creation of IMAX

During Expo 67 in Montreal, two shorts produced by the National Film Board of Canada (NFB) were presented that used 35mm and 70mm film projected simultaneously onto five screens, combining them into a single image in the manner of Cinerama. They were *In the Labyrinth* (Roman Kroitor, Colin Low, and Hugh O'Connor, 1967) and *Man and the Polar Regions* (Graeme Ferguson, 1967), and, according to *Time* magazine, they both produced a "stunning visual display […] [to] suggest that cinema—the most typical of 20th century arts—has just begun to explore its boundaries and possibilities."[101] But there were technical difficulties with the process that led Kroitor, Ferguson, Robert Kerr, and William C. Shaw to found the Multiscreen Corporation in September 1967 to design and build a single-camera/single-projector system that would produce large-screen images with a very tall aspect ratio (ultimately between 1.42:1 and 1.90:1) for theaters with steeply raked stadium seating. Their system used 70mm film running horizontally at 15 perforations per frame to produce an area nine times larger than standard 35mm and three times that of conventional 70mm film, which at the time ran vertically at five perforations per frame. Projected on a slightly curved screen more than seven stories tall (75–117 feet), such images provided a dramatically immersive experience. Two years later, the company was renamed IMAX, which is not an acronym but an invented word evocative of "maximum image." Appropriately Colin Low, one of the founders, later remarked "I always felt that IMAX should be in 3-D."[102]

In 1970, *Tiger Child* (Donald Brittain), the first IMAX film, was shown at Expo 70 in Osaka, Japan, and the following year the first permanent IMAX system was installed at the Cinesphere Theater in Toronto, debuting in May

101 "Magic in Montreal: The Films of Expo," *Time*, July 7, 1967.
102 Quoted in Zone, *3D Revolution*, 146.

1971 with the travelogue *North of Superior* (Graeme Ferguson). The first IMAX theater with hemispheric dome screen geometry was opened in 1973 in San Diego, where it doubled as a planetarium, and the first permanent IMAX 3D theater was built in Vancouver, British Columbia, in 1986 for Expo 86, where it showed the first IMAX 3D film, *Transitions* (Colin Low and Tony Ianzelo), produced by the NFB. This film is on the subject of transportation and features a continuously moving camera. Special visual effects artist Sean Phillips said of it, "Everything that comes after *Transitions* has to be viewed in a different light, because it really marked the dividing point between traditional 3-D, and what I call immersive 3-D."[103] (*Transitions* also contained the first stereoscopic computer animation, forecasting the immediate future of digital 3D.)

In addition to its use in purpose-build theaters, IMAX 3D was also used in theme park attractions and thrill rides, combined with motion simulation and other extra-stereoscopic effects like wind, smoke, scents, and watery mist to create an ever-greater immersive experience known as "4D." The first 4D commercial film was *The Sensorium* directed by Keith Melton and screened at the Six Flags Power Plant theme park in Baltimore in 1984, which used not IMAX but 35mm Arrivision over-and-under for its 3D effect. Subsequent 4D ride films directed by Melton were *Haunts of the Olde Country*, a haunted house experience installed at Busch Gardens in Williamsburg, Virginia, in 1993; *Pirates 4D*, a Caribbean adventure-comedy, installed at Anheuser-Busch's Sea World of Ohio in 1997; and *Tall Tales*, an attraction set in the southern swamplands designed for the Stone Mountain theme park outside of Atlanta, Georgia, in 2001, that contained some unprecedented off-screen CGI. All of Melton's films used a dual five-perforation 70mm system that ran vertically in camera and was less expensive than IMAX 3D, whose first ride film was *Race for Atlantis*, opening at Caesar's Palace in Las Vegas in 1998. This involved the audience in a simulated chariot race that was projected on a dome screen measuring 82 feet in diameter, the largest in the world at the time. (Other notable non-IMAX ride films were Iwerks Entertainment's *Dino Island II 3D* [1998] and Universal Studio's *The Amazing Adventures of Spiderman* [1999]; Disney's *Honey I Shrunk the Audience*, directed by Randal Kleiser [1994], and Universal's *T2 3-D: Battle Across Time* [1996], directed by James Cameron; these were short narrative entertainments that contained ride-like 4D elements, and all used some version of 15/75mm capture and projection.)

103 Ibid., 147.

IMAX 3D

Beyond theme parks and thrill rides, IMAX and IMAX 3D were initially (and still are) used mainly for documentaries of 40 minutes or less produced for institutional venues such as science centers and museums. The size and heft of the system's camera and projection equipment was prohibitive of long-form narrative. Representative early titles are *Grand Canyon: The Hidden Secrets* (Kieth Merrill, 1984), *The Dream Is Alive* (Graeme Ferguson, 1985; on NASA's space shuttle program), *On the Wing* (Bayley Silleck and Francis Thompson, 1986; a history of flight), *The Last Buffalo* (Roman Kroitor and Sally Dundas, 1990; IMAX 3D), *Ring of Fire* (George Casey, 1991; on volcanoes), *Hidden Hawaii* (Robert Hillman, 1993), *Destiny in Space* (Ben Burrt, 1994), *Africa: The Serengeti* (George Casey, 1994), and *The Living Sea* (Greg MacGillivray, 1995). Some of these were nominated for Best Documentary Short by the American Academy of Motion Picture Arts and Sciences. The first dramatic film to appear in IMAX 3D was the American-French coproduction *Wings of Courage* (Jean-Jacques Annaud, 1995), based on the exploits of aviation pioneer Antoine de Saint-Exupery; running just 40 minutes for the same reason as the documentaries, it was only nominally a feature. More short 3D narratives followed: *Across Time and Space* (Stephen Low, 1995), *T-Rex: Back to the Cretaceous* (Brett Leonard, 1998), *Sydney: Story of a City* (Bruce Beresford and Geoff Burton, 1999), and *Alien Adventure* (Ben Stassen, 1999)—technically, the first digital feature-length film produced in a large-screen format. (In 2000, Disney's *Fantasia 2000*, a sequel to its 1940 *Fantasia* that combined cel animation with CGI, became the first full-length animated feature to be released in IMAX, although it was not stereoscopic; the studio followed in 2002 with the animated science fiction adventure movie *Treasure Planet* [Ron Clements and John Musker], which was released simultaneously in regular and IMAX theaters.) IMAX 3D has continued be the main vehicle for IMAX documentary shorts, such as *Space Station 3D* (Toni Myers, 2002), *Bugs! A Rainforest Adventure* (Mike Slee, 2003), *NASCAR: The IMAX Experience* (Simon Wincer, 2004), *Magnificent Desolation: Walking the Moon in 3D* (Mark Cowen, 2005), *Deep Sea 3D* (Howard Hall, 2006), *Mummies 3D: Secrets of the Pharaohs* (Keith Melton, 2007; a 2D to 3D conversion), *Dinosaurs: Giants of Patagonia* (Marc Fafard, 2007), *Dinosaurs Alive! 3D* (David Clark and Bayley Silleck, 2007), *Sea Monsters: A Prehistoric Adventure* (Sean Phillips, 2007), *Grand Canyon Adventure: River at Risk* (Greg MacGillivray, 2008), *Wild Ocean* (Luke Creswell and Steve McNichols, 2008), *Dolphins and Whales 3D: Tribes of the Ocean* (Jean-Jacques Mantello, 2008), *Under the Sea* (Howard Hall, 2009), *Hubble* (Toni Myers, 2010), *Legends of Flight* (Steve Low, 2010), *The Last Reef* (Luke Cresswell and Steve McNichols, 2012), *To the Arctic 3D* (Greg MacGillivray, 2012), *Hidden Universe 3D* (Russell Scott, 2013),

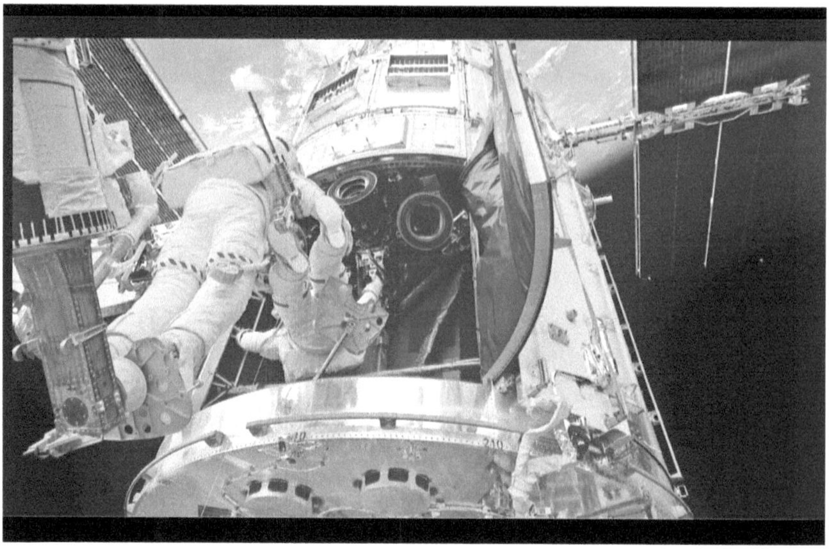

Figure 5.1 *Hubble* (Warner Bros. Pictures/IMAX, 2010): shot in 15/70mm IMAX 3D. Assembling the telescope. See Figure 6.6 for a dramatic reconstruction.

Journey to the South Pacific (Greg MacGillivray, 2013), *Wonders of the Arctic* (David Lickley, 2014), and *America Wild: National Parks Adventure* (Greg MacGillivray, 2016) (Figures 5.1 and 5.2).

The similarity of these titles says something important about IMAX 3D at this time—that it was deemed useful for exotic documentary spectacle but not much else. There were inherent limitations in the equipment—even when smaller 15/70mm camera systems were perfected for underwater and aerial photography, the projectors could only handle one large reel of film at a time, keeping running times to under 50 minutes (and usually closer to 40), except for special presentations like *Fantasia 2000*. This problem was partially solved by the advent of digital IMAX projection in 2008, designed for use with shorter 1.90:1 aspect ratios on smaller screens. The system initially used two 2K Xenon projectors to superimpose images to allow for brighter 2D images at a resolution of 2.9K and to display one image for each eye at a 2.5-inch offset for 3D presentations. In one 3D method, viewers wore polarized glasses to receive only the correct image for each eye; in another, the two projectors rapidly alternated frames (two or more at a time to enhance resolution) and viewers wore pseudo-helmets with liquid crystal shutters controlled by lasers in the theater space that blocked or transmitted light to the appropriate eye in synchronization with the projectors. Several early digital 3D films produced for release in conventional theaters were also

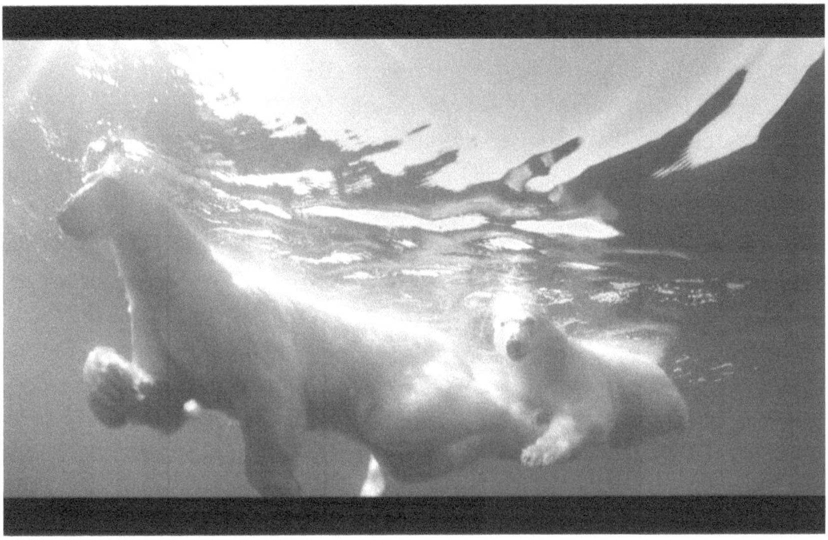

Figure 5.2 *To the Arctic 3D* (Warner Bros. Pictures/IMAX, 2012): shot in 15/70mm IMAX 3D. Swimming polar bears.

presented in digital IMAX 3D, notably *Avatar* (2009), *The Amazing Spider-Man* (2012), and *Gravity* (2013). Larger IMAX theaters, however, continued to use 15/70mm analog film until 2014, when the company rolled out its new Dual 4K laser projection system called "IMAX with Laser," replacing xenon arc lamps with laser beams that allowed digital projection on the full 1.43:1 surface of traditional IMAX screens. (IMAX with Laser was also capable of 60 fps and 50 percent greater brightness than standard IMAX digital projectors, and it required the use of dichroic filter glasses, which selectively pass light of a small range of colors while reflecting others, rather than linearly polarized glasses.)[104]

In 2008, however, most IMAX 3D cameras still used film and weighed over 249 lbs; their technical limitations in terms of producing standard-length features included their size and weight, noise (which inhibited dialogue recording), short 70mm film magazines, and the greater cost of film stock. In 2014, IMAX introduced the Phantom 65 Gold, a 38-lb 4K digital 3D camera used to film 94 minutes of special sequences in Michael Bay's 164-minute-long

104 Dichroic filters produce highly saturated colors in the same way as oil films on water. When light strikes an oil film at an angle, some of the light is reflected from the top surface of the oil, and some is reflected from the bottom surface where it is in contact with the water.

Transformers: Age of Extinction. It featured a custom 4K horizontal sensor that was physically larger than any other digital camera sensor and could capture more information, particularly a better range of dark and light detail, producing a frame about 4,000 pixels wide. IMAX supplied a set of fixed-focal length lenses and also used a dual digital engine to control convergence, so that the camera could shoot 3D images at full 4K resolution at 100 fps. *Age of Extinction* thus became the longest IMAX 3D movie ever filmed, and it was followed by *Gone with the Bullets* (Wen Jiang, 2014), a Chinese 3D feature that also used the Phantom 65 Gold for some of its most spectacular sequences. Yet IMAX only gradually phased out its 15/70mm film cameras, and they were used for a number of 2D and 3D films of the period, for example, *The Dark Knight* (2008), *Transformers: Revenge of the Fallen* (2009), *Mission Impossible: Ghost Protocol* (2011), *The Dark Knight Rises* (2012), *Star Trek into Darkness* (2013; converted to 3D), *The Hunger Games: Catching Fire* (2013), *Interstellar* (2014), *Star Wars: The Force Awakens* (2015; converted to 3D), *Batman v. Superman: Dawn of Justice* (2016; converted to 3D), *Dunkirk* (2017), *Star Wars: The Last Jedi* (2017; converted to 3D), and *First Man* (2018). In 2016 IMAX joined with the Arriflex Corporation to produce the Arri Alexa IMAX digital camera, and several 3D films were shot in whole or part with that rig, notably *Captain America: Civil War* (2016), *Transformers: The Last Knight* (2017), *Avengers: Infinity War* (2018), and *Avengers: Endgame* (2019). (Besides the IMAX Phantom 65 and the Arri Alexa, there were several other digital camera systems that sustained post-*Avatar* 3D: the RED Dragon, the RED Epic, and the Canon EO, with many more to follow.)

A landmark in the history of 15/70mm IMAX 3D was *The Polar Express* (Robert Zemeckis, 2004), a computer-animated adventure film based on Chris Van Allsburg's 1985 children's book of the same title; it used performance-capture technology combined with three-dimensional CGI to tell the story of a young boy's magical trip to the North Pole on Christmas Eve. The left- and right-eye animation files were composted in 2K resolution by a virtual camera[105] and converted to 4K by the IMAX DMR (digital remastering) process, then printed onto 65mm IMAX negative stock that was in turn printed onto IMAX 15/70mm positive stock for distribution

105 According to Miriam Ross, *The Polar Express's* production team was able to "photograph" corresponding left- and right-eye images using a mathematical formula to represent different cameras: "These virtual cameras were able to calculate the desired focal length of the virtual lens and animate a virtual interaxial distance so that a smooth progress between different depth planes was possible" (Ross, *3D Cinema* [London: Palgrave Macmillan, 2015], 177).

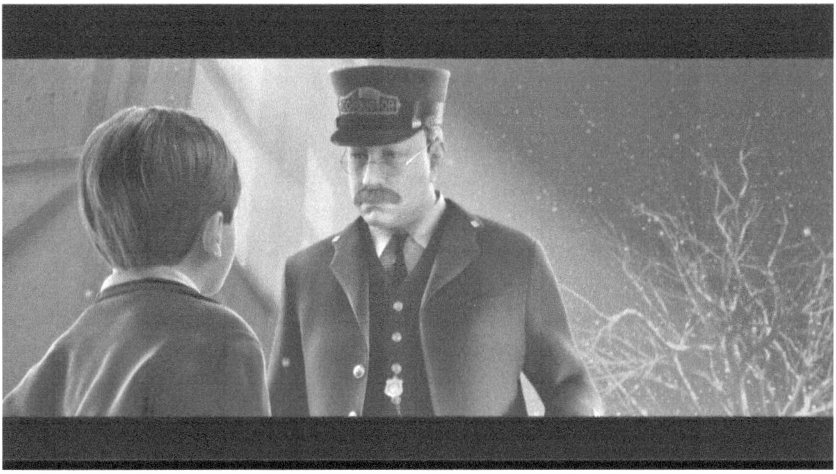

Figure 5.3 *The Polar Express* (Warner Bros. Pictures/IMAX, 2004): one of the first features shot in 15/70mm IMAX Digital 3D. The left- and right-eye animation files were composted in 2K resolution by a virtual camera and converted to 4K by the IMAX DMR (digital remastering) process, then printed onto 65mm IMAX negative stock, which was in turn printed onto IMAX 15/70mm positive stock for distribution. Performance capture technology creates an "uncanny valley" effect in the conductor's (Tom Hanks's) eyes. NB: After IMAX Digital 3D, assume digital sound recording and reproduction for all digital stereoscopic films—variants of Dolby Digital, Dolby Atmos, DTS, or SDDS. Also assume that all features are processed by Technicolor Laboratories, and digitally color-graded, unless otherwise noted.

to nearly 70 large-format theaters in the United States and Canada (Figure 5.3). (With the advent of the DMR blowup process, IMAX projectors were modified to provide a run time of up to 175 minutes, making possible the exhibition of full-length features such as *The Polar Express*, later *Beowulf*, and *Avatar*; in digital projection formats, of course, length was not a problem.) *The Polar Express* was also given wide release in flat 35mm anamorphic format, and it became a box-office hit, grossing over $300 million worldwide against its $165 million budget. Following this trend, the first 35mm digital 3D films tended to be animated ones, such as DreamWorks' *Monsters vs. Aliens* (Conrad Vernon, 2009) and Disney/Pixar's *Cars 2* (Brad Lewis, 2011). (Both had limited IMAX 3D releases as well.)

As IMAX 3D supervisor Hugh Murray said at the time, "CGI is almost the ideal medium for 3-D."[106] There are several reasons for this. For one thing,

106 Quoted in Zone, *3D Revolution*, 236.

depth of field, so important to left- and right-eye convergence, can be artificially manipulated in CGI or even eliminated since the "focal length" of a virtual camera lens is simply a mathematical equation, and the interaxial distance between left- and right-eye images can be variably altered from large closeups to long-distance perspectives to suit audience point of view. Furthermore, computer-animated films rely on ray-tracing software and volumetric distortion to create the illusion of three-dimensionality, so that their digital files can be easily offset for 3D presentation. In other words, computer-animated 2D films made with 3D models can be rendered stereoscopic by effectively adding a second virtual camera, which is technically not a conversion since such reconstructions have the same three-dimensional quality as films produced in native 3D. The rerelease of *Toy Story* (John Lassiter, 1995) and *Toy Story 2* (John Lassiter, 1999) in 2009 is a good example of this phenomenon; many other flat computer-animated films produced with 3D wireframes would follow suit later in the decade, for example, the 3D rerelease of *The Nightmare before Christmas* (Tim Burton, 1993) in 2007, 2008, and 2009.

At the same time, some live-action and conventional cel-animated films were undergoing actual 2D to 3D conversion. For example, the large-format *Mummies: Secrets of the Pharaohs* (Keith Melton, 2007) was originated on flat 15/70mm and converted to IMAX 3D in 2008 through the use of digital postproduction tools. Such conversions are often done to increase a film's marketability in new venues. The distinction between a 2D film converted to digital 3D and one originally conceived for stereoscopic viewing is that the former has no three-dimensional depth cues from the moment of acquisition. It is not authored in 3D and can therefore be a very unsatisfying experience in a theater (e.g., Disney's 2011 3D IMAX conversion of *Beauty and the Beast* [1991], in which the studio invested less than $10 million). The most important distinction between them lies in the presence of a 3D camera, such as the Arri Alexa or Cameron Pace Fusion, whose presence is a guarantee that a given film is not a 3D conversion. Yet in movies such as Rob Minkoff's *The Lion King* (1994; limited IMAX 2D rerelease 2002; 3D conversion 2011), Tony Scott's *Top Gun* (1986; 3D conversion and limited IMAX 3D release 2013), Steven Spielberg's *Jurassic Park* (1993; 3D conversion 2013), and James Cameron's *Titanic* (1997; 3D conversion and limited IMAX 3D release 2012) and *Terminator 2: Judgment Day* (1991; 3D conversion 2017), where the conversion is being supervised by the film's original author and where the depth cues are manifest because of the subjective nature of the process, conversions can be just as satisfying as films scripted and storyboarded for three dimensions. A recent example of an exquisite 2D to 3D conversion is Warner Bros.'s 2013 IMAX 3D restoration and 75th Anniversary Blu-ray edition of *The Wizard of Oz* (1939), initiated from a very high-resolution 8K scan of the original Technicolor negative and then creating a depth map of each 2D frame to construct 3D imagery and determine the distances from the viewer's point of view with the use of a rotoscope.

RealD 3D

In 2005, a California company called RealD was established to market a new digital stereoscopic projection process that used circularly polarized light to produce three-dimensional images; its advantage over linear polarization was that viewers could tilt their heads without seeing doubled or darkened frames. Developed for use in theaters with high-resolution video projectors, either Texas Instrument's DLP (digital light processing) devices or Sony's reflective LCOS (liquid crystal on silicon) units, the RealD system switches between left- and right-eye frames 144 times per second. (Each frame is projected three times to reduce flicker, a practice known as triple flash.) Over the next decade, RealD became the most widely used technology for the theatrical projection of 3D movies, with more than 27,000 installations in 72 countries by mid-2015, compared to IMAX's 1,000. Nevertheless, RealD had significant problems with screen brightness early on, which will be addressed in subsequent chapters.

In 2005, *Chicken Little* (Mark Dindal) became Disney's first 3D computer-animated film and its first to be released in 3D since *Melody* (1953; see Chapter 3). Initially intended for 2D, it was converted to 3D during production for release in the new RealD single-projector polarized exhibition process. Before that happened there was a last gasp of big-budget anaglyphic stereoscopy in Robert Rodriguez's *Spy Kids 3-D: Game Over* (2003) and *The Adventures of Shark Boy and Lava Girl in 3-D* (2005). Both were shot with the same Reality Camera System built by James Cameron and Vince Pace for Cameron's IMAX 3D documentaries *Ghosts of the Abyss* (2003) and *Aliens of the Deep* (2005), then postconverted into polychromatic 3D for exhibition in a way that one critic said caused "protracted retinal strain."[107] (The last commercial gasp of anaglyphic was probably Tom Savani's remake *Night of the Living Dead 3D*, released in both anaglyphic 35mm and D-cinema frame-sequential 3D in 2006.) These anomalies notwithstanding, Robert Zemeckis's ImageMovers production company turned to 3D performance capture again for the family-friendly haunted house movie *Monster House 3-D* (2006), working with the same

107 Quoted in Zone, *3D Revolution*, 254.

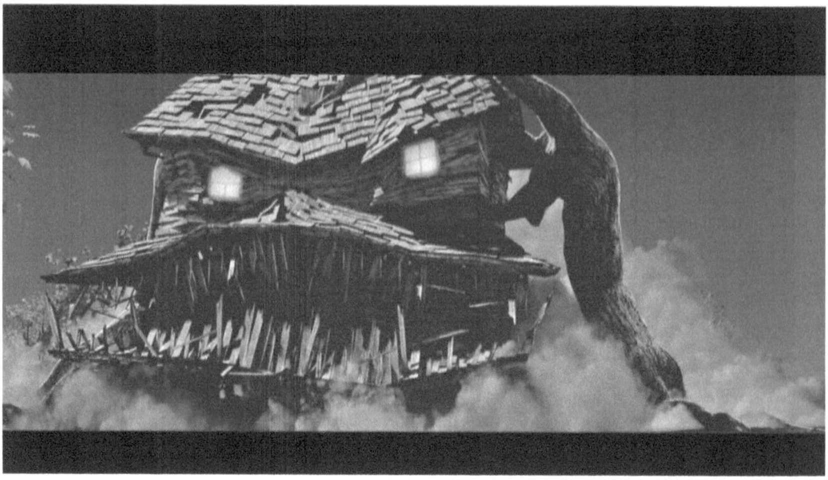

Figure 5.4 *Monster House 3-D* (Sony Pictures Releasing, 2006): the first computer-animated digital stereoscopic features to use performance capture, presented in RealD Cinema. The haunted monster house in question.

Sony Imageworks team that had produced the IMAX 3D version of *The Polar Express* in 2004 (Figure 5.4). Together with *Chicken Little 3-D*, *Monster House* was an important catalyst for the rollout of digital cinema generally: the former opened on 84 screens in November 2005, and the latter opened on twice that number just eight months later (although, it was ultimately not as popular).[108] This was significant: stereoscopy was now poised to lead the mainstream cinema for the first time since 1952.

Zemeckis followed with a 3D performance-capture feature based on the Old English epic poem *Beowulf* (2007) that was realistic rather than fantastic in mode. Using 244 motion-capture cameras, as opposed to 67 for *The Polar Express*, Zemeckis was able to achieve shots impossible for straight live action. In this digitally enhanced filmmaking process, actors on set wore 78 body markers and physical cameras shot real-time footage of their work. The director then uses a virtual camera to frame the action, while two teams of computer animators labored to refine facial expressions and body movement. The performance capture in *Beowulf* was considerably more sophisticated than that of *The Polar Express*, and it largely avoided the "uncanny valley" effect of character representation that marred its predecessor.

108 Whereas *Chicken Little* earned $342,432,837 against its $50 million budget, *Monster House* barely returned its $65 million costs.

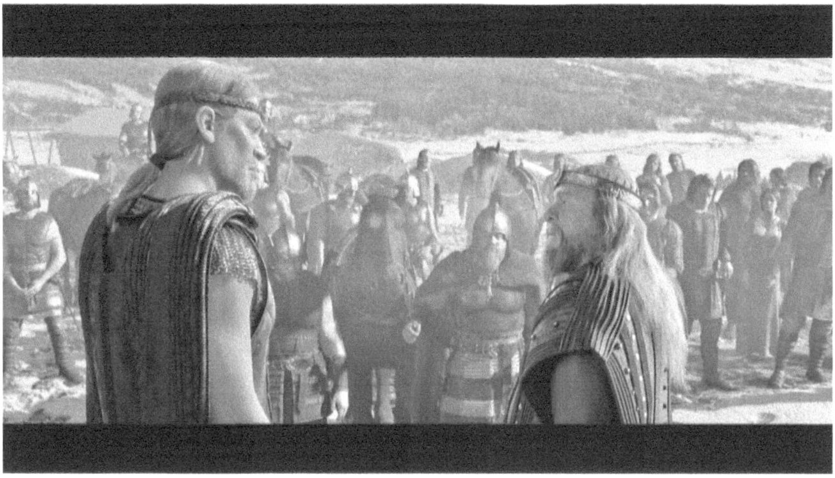

Figure 5.5 *Beowulf* (Paramount Pictures/Warner Bros. Pictures, 2007): shot in Digital 3-D anamorphic (2.39:1) for presentation in RealD Cinema and blown up to 70mm for IMAX. The legendary warrior Beowulf (Ray Winstone) meet King Hrothgar of Norway (Anthony Hopkins) in performance captured images.

(Figure 5.5).[109] About the quality of the 3D, however, there was no question. *Rolling Stone* critic Pater Travers wrote, "The eighth-century *Beowulf* [...] will have you jumping out of your skin and begging for more [...] I've never seen a 3-D movie pop with this kind of clarity and oomph." And in *Empire*, Tom Ambrose claimed that the film was the finest example of the capabilities of digital 3D yet: "Previously, 3D movies were blurry, migraine-inducing affairs [...] *Beowulf* it a huge step forward." Be that as it may, Paramount spent $150 million to produce the film, and it disappointed commercially, earning just $195,394,745 worldwide.

Also notable at this time were Disney's *Meet the Robinsons* (Stephen Anderson, 2007), a computer-animated science fiction comedy released in both standard 35mm and Disney Digital 3-D (a brand name for Disney films projected using any digital 3D technology, in this case RealD), and *Scar* (Jed Weintrob, 2007), a nasty torture porn venture that was, not incidentally, the

109 A term used in aesthetics and robotics, the "uncanny valley" concept suggests that humanoid objects that imperfectly resemble real humans evoke feelings of revulsion in observers, particularly with regard to the representation of eyes, which are frequently described as "eerie" or "creepy." NB: Justin Chang's *Variety* review of *Beowulf*: "While the technology [of performance capture] has improved since 2004's *The Polar Express* (particularly in the characters' more lifelike eyes), the actors still don't seem entirely there."

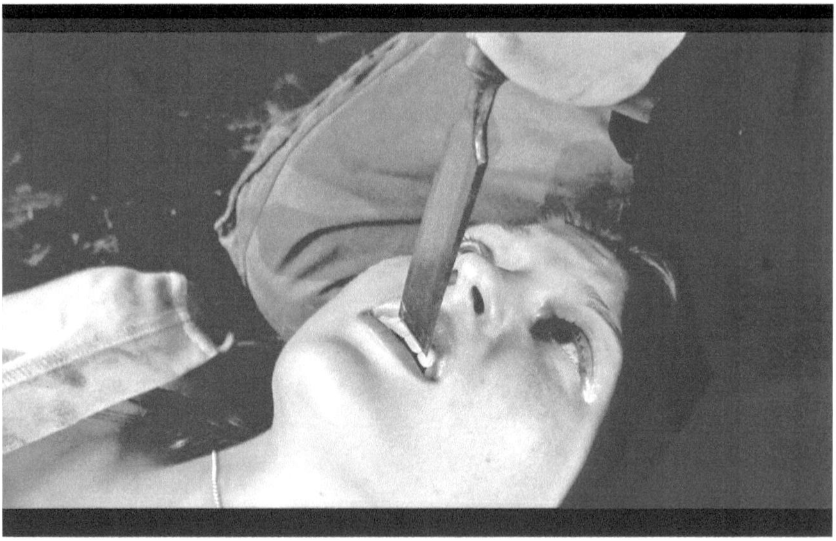

Figure 5.6 *Scar* (Phase 4 Films, 2007): the first feature to be shot in HD 3D (1.78:1) and to be completed in a totally digital postproduction environment. Torture porn in three dimensions.

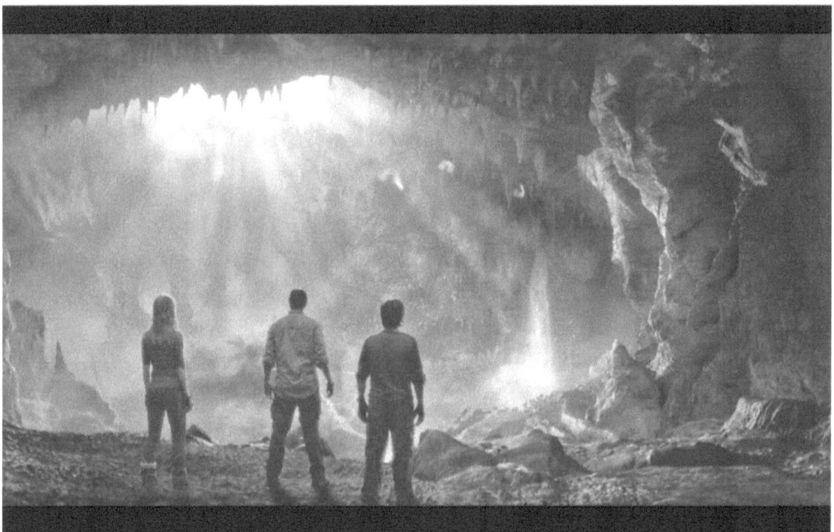

Figure 5.7 *Journey to the Center of the Earth* (New Line Cinema/Warner Bros. Pictures, 2008): shot in HD 3D (1.78:1) for presentation in RealD Cinema. Wonder at the earth's center.

first feature to be shot in HD 3D and to be completed in a totally digital postproduction environment (it also became, in 2010, the first Video on demand movie to be streamed for 3D televisions by cable) (Figure 5.6). Altogether more respectable was the Jules Verne adaptation *Journey to the Center of the Earth* (Eric Brevig, 2008), which introduced the 4DX format in its South Korean debut, using motion simulation, water, light, and fog effects in the theater space (Figure 5.7). This film was a critical and popular success, earning $242 worldwide on a $60 million budget, paving the way for the watershed year of 2009, as did the Belgian-American computer-animated science fiction comedy *Fly Me to the Moon* (Ben Stassen, 2008), released in standard digital and IMAX 3D, and Disney's *Bolt* (Chris Williams and Byron Howard, 2008), the first computer-animated film to feature non-photorealistic 3D rendering.

2009

The year 2009 was an auspicious one for three-dimensional cinema, ending as it did with the release of James Cameron's *Avatar* in December, which went on for a decade to become the highest grossing film in the history of motion pictures. Public acceptance of 3D, driven by the popularity of IMAX documentaries and computer-animated features, was high, despite the higher ticket upcharge that most exhibitors levied on the process. January saw the wide release of *My Bloody Valentine 3D* (Patrick Lussier), a remake of the 1981 slasher and the first R-rated film to be projected in RealD. A modest critical success, its 3D sequences grabbed attention, with Frank Scheck writing in *The Hollywood Reporter*, "While the concept of adding 3-D to the horror genre is hardly new, Patrick Lussier's film is the most accomplished example," harking back to the *frisson* of *House of Wax* (Figure 5.8). In February, Focus Features released *Coraline* (Henry Selick, 2009), a 3D stop-motion animation fantasy about a young girl who discovers a sinister alternative universe within her own house. That the concert film could be a successful 3D genre had already been demonstrated by National Geographic's *U2 3D* (Catherine Owens and Mark Pellington) and Disney's *Hannah Montana and Miley Cyrus: The Best of Both Worlds* (Brice Smith), both of which enjoyed significant box-office success in 2008, a trend continued the following year by Disney's *The Jonas Brothers: The 3D Experience* (Bruce Hendricks).

Dreamworks Animation entered the field in March 2009 with *Monsters vs. Aliens* (Conrad Vernon and Rob Letterman), its first feature to be directly produced in stereoscopic format rather than converted from 2D after the fact. This parodic, intertextual mélange of classic monster and science fiction movies grossed over $321 million against its $174 million budget, suggesting

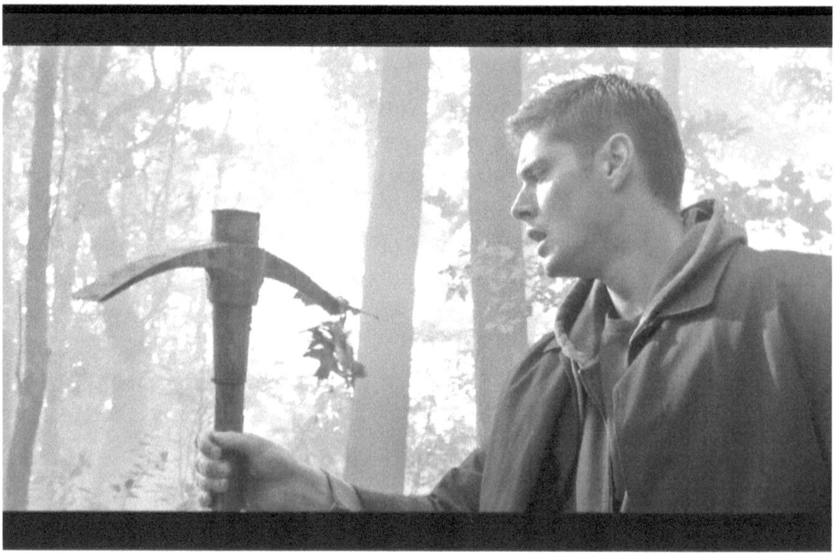

Figure 5.8 *My Bloody Valentine 3D* (Lionsgate, 2009): shot with a Red One Digital 3D camera system (1.85:1) for presentation in RealD Cinema. The mentally disturbed killer (Jensen Ackles) finds his tools.

sustained interest in the 3D process (In fact, after the success of *Monsters vs. Aliens*, Dreamworks vowed to release all of its subsequent films in 3D.) (Figure 5.9). Two other popular computer-animated comedy adventure films appeared in 3D in May and July of 2009 respectively—Pixar/Disney's *Up* (Pete Docter), about a grumpy old widower who escapes his loneliness by tying thousands of balloons to his house and sailing into the wilds of the Venezuelan jungle, and Fox's *Ice Age: Dawn of Dinosaurs*, in which a herd of prehistoric animals has a series of raucous interactions with other denizens of the frozen veld (Figure 5.10). *Up* was a great critical success, becoming the first animated film to open the Cannes Film Festival and a five-category Oscar nominee (winning for Best Animated Feature and Best Musical Score); it was also the fifth highest grossing film of 2009 worldwide, while *Ice Age* was the third. In August, *The Final Destination 3-D* (David R. Ellis), the fourth entry in the supernatural horror franchise and the first in HD 3D, debuted at the top of the North American box office, where it remained No. 2 for weeks, ultimately grossing $186.5 worldwide. (*Final Destination 4*, as it was also known, was the first 3D film to offer D-BOX technology in selected US theaters, although it had been used the year before in the South Korean release of *Journey to the Center of the Earth*; this is a system that synchronizes motion-simulation effects with visual content, providing

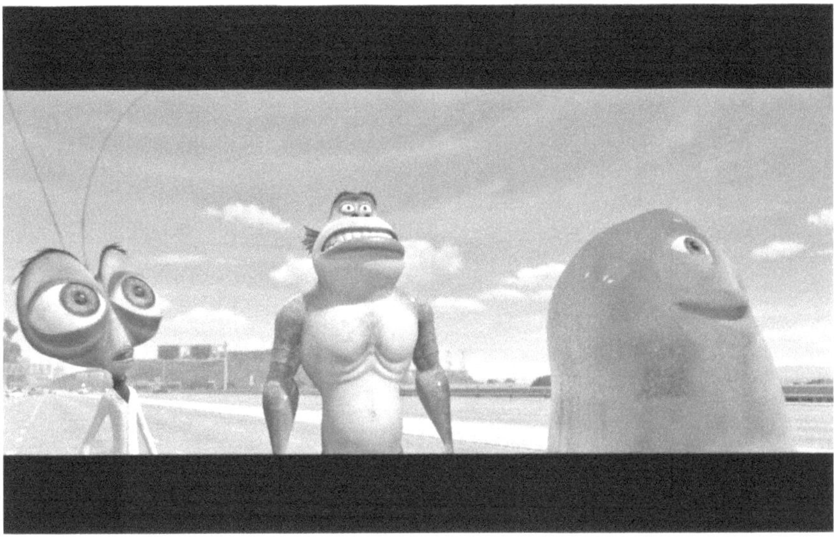

Figure 5.9 *Monsters vs. Aliens* (Dreamworks Animation/Paramount Pictures, 2009): shot in Digital 3-D anamorphic (2.39:1) for presentation in RealD Cinema. Three computer-animated monsters.

a partial 4D experience.) Several other live-action 3D genre films appeared in 2009 with less box-office appeal, for example, *Call of the Wild* (Richard Gabai), an updating of Jack London's famous story shot on location in Montana, and *Dark Country* (Thomas James), a noirish psychological thriller set in the Southwestern desert country of Nevada and New Mexico. More financially successful stereoscopic films were Sony's computer-animated science fiction comedy *Cloudy with a Chance of Meatballs* (Phil Lord and Christopher Miller) and Disney's motion-capture fantasy *A Christmas Carol* (Robert Zemckis), which one critic described as "an exasperating re-imagining of the Dickens classic as a 3-D action-thriller zooming through Victorian London [...] [with the moral] [...] lost in the snowdrifts of special effects."[110]

James Cameron has been planning the film that became *Avatar* for years before it went into production in February 2007. Cameron's experience with 3D began with *T2 3-D: Battle Across Time* (1996), a twin-strip 70mm theme park attraction for Universal Studios, and the IMAX 3D documentaries *Ghosts of the Abyss* (2003) and *Aliens of the Deep* (2005), both of which he had shot with the twin Fusion Camera HD 3-D System

110 Zone, *3D Revolution*, 329.

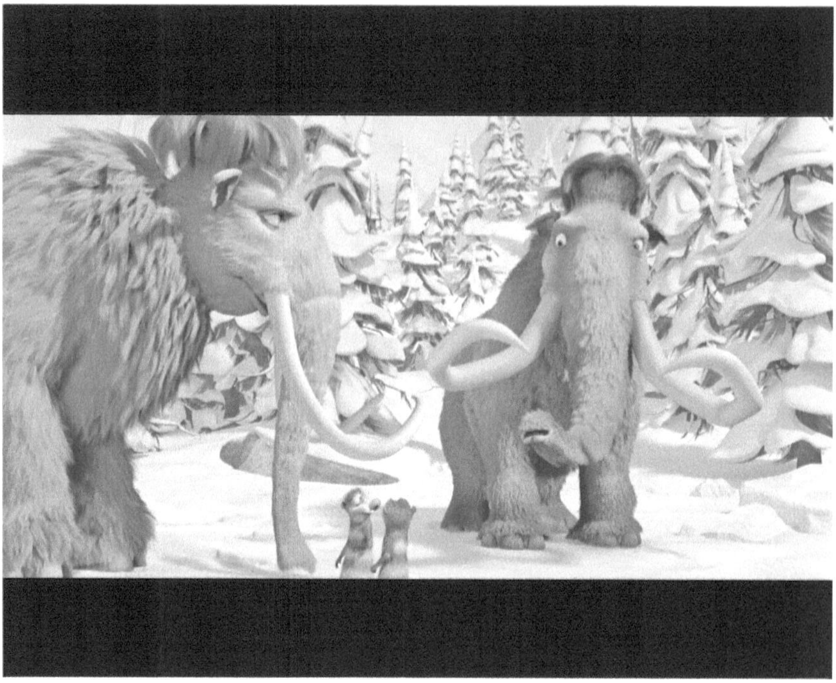

Figure 5.10 *Ice Age: Dawn of the Dinosaurs* (20th Century Fox, 2009): shot in Digital 3-D at 1:85 for presentation in RealD Cinema. Some computer-animated mastodons.

(aka Reality Camera System, or RCS), that he and Vince Pace built specifically for underwater filming. For the principal photography of *Avatar*, Cameron and Pace modified the Fusion rig so that it could shoot full live action in combination with CGI characters and real locations, combining two Sony F-950 cameras and a beam-splitter, together with a new stereo camera technology they called the Simulcam, which was used to integrate performance capture of the actors into a computer-generated fantasy world created by Peter Jackson's Weta Digital effects studio in New Zealand (40 percent of the movie was live action; 60 percent was CGI). This place was Pandora, a lushly forested moon in the Alpha Centuri galaxy inhabited by an alien species called Na'vi (Hebrew for "prophet"), 10-foot tall, blue-skinned humanoids who worship nature. Cameron became convinced that CGI had progressed enough to sustain his vision when he saw the motion-capture presentation of Gollum in Peter Jackson's *The Lord of the Rings: The Two Towers* (2002) (Figure 5.12).

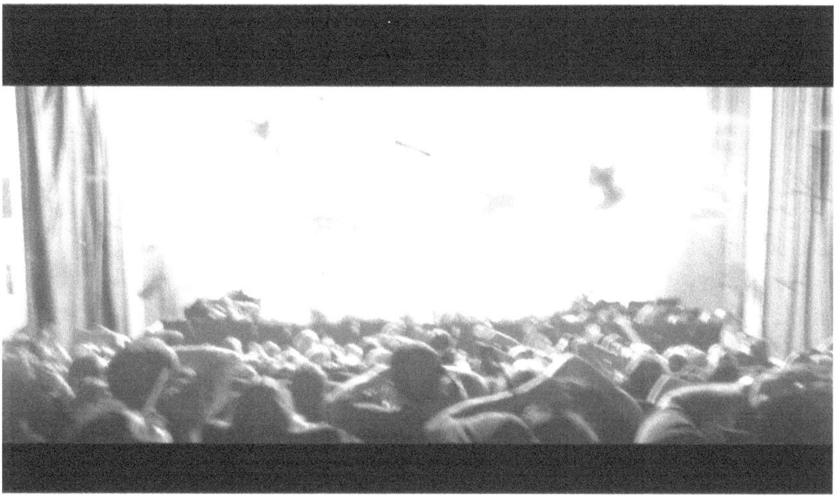

Figure 5.11 *The Final Destination 3D* (New Line Cinema/Warner Bros. Pictures, 2009): shot with Sony's Cinealta Digital 3D system (2.39:1) for presentation in RealD Cinema. A spectacular meta-joke: a 3D screen in the movie explodes catastrophically into the audience for both films.

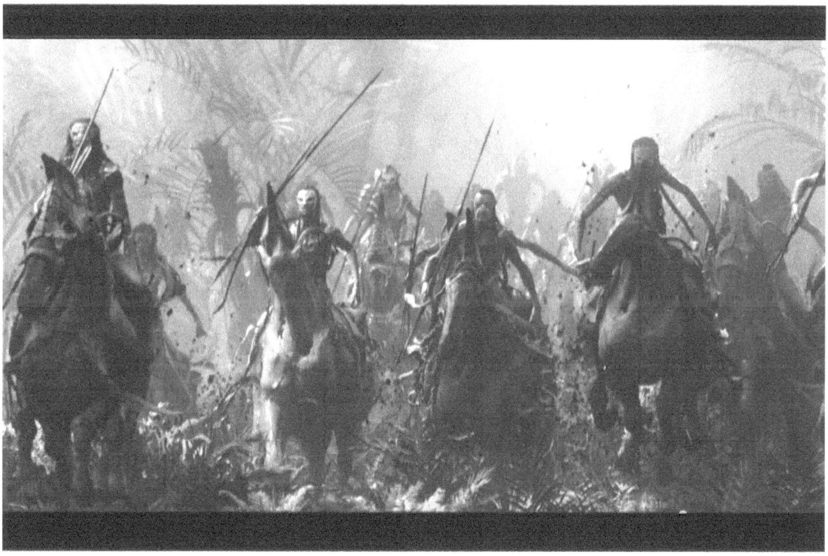

Figure 5.12 *Avatar* (20th Century Fox, 2009): shot in PACE Fusion 3-D (Simulcam) for presentation in IMAX Digital 3D (1.85:1) and RealD Cinema (2.39:1). The Na'vi fight back against an American invasion: Sam Worthington as Jake Sully in Na'vi form.

Avatar itself is a science fiction epic set in the mid-twenty-second century when humans, having depleted the energy resources of Earth, are attempting to plunder Pandora in order to extract its supply of the valuable mineral "unobtanuim." Jake Sully is the eponymous "avatar" (which means "incarnation" in Sanskrit), a paraplegic former Marine who operates a genetically matched Na-vi–human hybrid as part of a scientific mission to gather information about Pandora's ecology and society. In somewhat conventional fashion—think *Dance with Wolves* (Kevin Costner, 1990), confirmed by Cameron as one of *Avatar*'s models, and Disney's *Pocahontas* (Mike Gabriel and Eric Golberg, 1995), from which Cameron says he got the idea—Jake falls in love with a native Pandoran named Neytiri and is converted to the Na'vi cause to resist the aggression of his militaristic employers, pushing all of the right liberal Hollywood buttons. Despite the mundanity of its plot, described in a press release as "an emotional journey of redemption and revolution," the three-dimensional world created by *Avatar* was unprecedented. The film had two different production designs, one for the forests of Pandora and the other for the human world, and the Simulcam system flawlessly integrated both with eye-catching parallax values. Cameron understood that overall parallax could be minimized in a digital environment to reduce the amount of optical work an audience needed to create convergence and make *Avatar* pleasant to view over its 161-minute length on gigantic IMAX 3D screens. As Kenneth Turan wrote in his December 17 review in the *Los Angeles Times*, "A Dazzling Revelation": "Think of *Avatar* as *The Jazz Singer* of 3D filmmaking. [...] [With it] he [Cameron] restores a since of wonder to the moviegoing experience that has been missing for far too long."[111] Even Roger Ebert, a perennial foe of 3D, said, "Cameron's iteration is the best I've seen—and, more importantly, one of the most carefully employed."[112]

Produced over four years at a cost of $237 million, *Avatar* opened in North America on December 15, 2009, at more than 2,100 3D and 1,200 2D venues, and its popular success is now legendary: with worldwide box-office sales of $2.79 billion it remained the highest grossing film in history until 2019, when it was nudged out by *Avengers: Endgame.* Nominated for nine Academy Awards, including Best Picture (the first movie shot with a 3D camera, released in 3D and IMAX 3D, to do so) and Best Director, it won for Best Art Direction, Best Cinematography (the first completely digitally shot film to do so), and Best Visual Effects. Despite some critical carping about its storytelling and its "white savior" theme, industry analysts recognized *Avatar* as the holy grail of

111 Quoted in ibid., 389.
112 Ibid.

3-D, one of them explaining "This is why all these 3-D venues were built." And it was true—those venues would soon be pumped full of new potential blockbuster product, motivating the long-debated conversion from analog to digital projection for both 2D and 3D in theaters around the world. The die was cast: In Hollywood, it was like 1952 all over again, except that now digital technology offered the promise of three-dimensional cinema without clumsy, redundant equipment and without eye-straining parallax distortion.

Chapter 6

THE BLOCKBUSTER YEARS: DIGITAL 3D, 2010–20

The blockbuster syndrome

During the 1970s, the calculus by which a movie was considered successful or not in the marketplace had changed. Between the mid- and late 1960s, the studios teetered on the brink of insolvency because of the impact of the Paramount decrees, the national diffusion of television, and poor financial decisions made in their wake—the abortive attempt to convert to full stereoscopic production in the early 1950s among them. Starting with Universal's acquisition by MCA in 1962 and Paramount's absorption by Gulf and Western Industries in 1966, the classical Hollywood studios were all bought up by larger corporations who didn't really know how to run them, resulting in an industry-wide recession between 1969 and 1971. At the same time, in October 1968, the 34-year-old Production Code Administration (PCA) was replaced by a Code and Ratings Administration (CARA) that transformed the national audience—more or less homogeneous since its formation, except at the ethnic fringes—into a demographically segregated one. The conviction grew in Hollywood that the regular flow of product was no longer profitable because there was no longer a regular and predictable audience for its films. In such a volatile context, if a few productions became big hits, they could amortize those that produced a loss or simply broke even. This idea of engineering "event" movies to generate windfall profits and carry the rest along with them became known as "the blockbuster syndrome." This was not a new phenomenon—"blockbusters" like Paramount's *The Ten Commandments* (Cecil B. DeMille, 1956) and MGM's *Ben-Hur* (William Wyler, 1959) had kept their respective studios profitable for years after their release, and expensive flops like *Cleopatra* (Joseph Mankiewicz, 1963) had nearly ruined 20th Century Fox. What was different was the amount of capital involved in hedging the bets.[113]

---

113 *The Graduate*, for example, the top-grossing film of 1967, was shot for $3 million and made $104.4 million domestically; *Bonnie and Clyde* was shot in the same year for

Between 1972 and 1979, the average production budget or "negative cost" for a single film rose by 178 percent, or nearly four times the rate of general inflation. By the end of 1979, negative costs had doubled their 1977 level to reach the then-staggering sum of $8.9 million, so that the financial risks of production were substantially multiplied. This caused a trend toward the production of fewer and fewer films, plus a steady increase in the amount spent on advertising, marketing, and promotion to ensure a hit, with these expenses often rising more than twice as high as production costs. Thus the new blockbuster calculus: at the beginning of the 1980s, a film had to recoup three dollars at the box office for every dollar spent on production before it could turn a profit. Major ingredients of the blockbuster strategy were the adaptation of "presold" properties (e.g., best-selling novels and comic books) that would have built-in audience recognition, saturation booking (opening a film "wide" at thousands of theaters simultaneously), and national mass-media marketing campaigns (at the time, largely through expensive prime-time television advertising). Films like Paramount's *The Godfather* (1972), Warner's *The Exorcist* (1974), Universal's *Jaws* (1975), and Fox's *Star Wars* (1977) could ensure a studio's bottom line for years to come, but a single big-budget failure might threaten disaster. In the four cases noted here, the negative costs to gross earnings ratios in millions of dollars were, respectively, 7:86, 9:89, 12:130, and 9.5:262, and, additionally, the now legendary *Star Wars* merchandizing bonanza was worth over $4 billion to Fox/Lucasfilm going forward. In fact, during the 1980s and 1990s, film series like the Star Wars trilogy and its reboots, Paramount's Star Trek and Warner Bros.' Batman, all became huge product franchises whose branded merchandise was worth billions more in excess of box-office receipts.[114]

And yet between 1990 and 1995, studio marketing costs rose by 92 percent, star salaries doubled, and production costs experienced double-digit growth, at the same time that box-office revenues increased by only 9 percent. Concurrently, the average profit margin for features shrank to 14–15 percent. This dramatic inflation of costs threatened the industry in a major way, as did the number of films released, and in 1995 an industry financial analyst

---

$2.5 million and made $50.7 domestically, astonishing rates of return, but the windfall profits of 1970s films like *The Godfather*, *The Exorcist*, *Jaws*, and especially *Star Wars* (the decade's most stunning sleeper, produced for $11 million and earning $775.8 million) dwarfed these numbers, even during a period of rampant monetary inflation.

114 By the 1990s, merchandizing, which began modestly in the early twentieth century (Mickey Mouse watches, Shirley Temple dolls, etc.), had risen from the category of ancillary income to become a significant studio revenue stream, and all of the majors operated large consumer product divisions.

wrote, "This is the worst time to be producing motion pictures in the history of the business." By that time, only 35 percent of industry earnings came from theatrical release (both domestic and foreign, the latter now exceeding the former), while the remainder came from ancillary markets such as merchandising and video licensing. Wide releases became the new normal, so that today it is not uncommon for a major studio release to open on 2,000 or even 3,000 screens at the same time. Such wide openings put an inordinate emphasis on the three-day grosses for a film's opening weekend, which are then used to set the values for the film in all of its subsequent windows—cable, videodisc, and streaming. In 2007, the MPPA calculated that the average production budget for a standard A-list feature had risen to $65 million, exclusive of marketing and distribution costs; seven years later, *The Hollywood Reporter* noted that negative costs were averaging $100 million per major feature, with attendant international marketing costs crossing the $200 million mark. In this context, it isn't hard to see why the nearly $2.79 billion gross of *Avatar* in 2009 appealed to a now institutionalized blockbuster mentality in Hollywood, causing a contagious new wave of stereoscopic fever.

3D television

Somewhat unexpectedly, the most immediate impact of the *Avatar* phenomenon was felt in the electronics industry: in 2010, all of the world's major television manufacturers, starting with Panasonic, Sony, and LG, began to roll out 3D-capable models. As the size of flat-screen TVs had grown and their prices had dropped—and new models had all become digital—the cost of adding 3D capability was relatively modest, perhaps several hundred dollars per unit. Doubling and offsetting digital video images as opposed to analog ones was not a complicated matter; in fact, several companies offered sets that could convincingly convert standard 2D digital content into 3D—a boon to watching certain sporting events or 2D movies composed in depth. Indeed, it may be true to say that the post-*Avatar* 3D revolution was at first driven by the home theater market more than theatrical exhibition, the irony being that the second has apparently killed the first (and, in some ways, vice versa).

As in theatrical venues, there are two ways to view home media 3D images. Active-shutter systems use LCD lenses that are wirelessly synchronized to the theatrical film or TV frame rate: when the image intended for the left eye is shown, the lens for the right eye goes dark, and vice versa. Because active-shutter glasses require batteries, they are relatively bulky and expensive, but they create the most solid kind of 3D effects and are standard equipment for IMAX 3D presentations; they have rapid refresh rates, resulting in full-resolution images for both left and right eyes. On the downside, shutters produce dimmer and sometimes flickering images. The second type, passive 3D systems, uses polarized lenses as explained above: each eye's respective image is filtered in such a way that the lens for the opposite eye blocks it from view, resulting in a resolution half that of 2D or active-shutter glasses, but with brighter images and less crosstalk. RealD theaters use passive glasses because they are cheaper, lighter, and will work with any passive polarized projection. Active and passive systems are incompatible in both in the theater and the home, and different TV manufacturers dedicated their products to one or the other so that brands were not interchangeable. (Active systems were common to Mitsubishi, Panasonic, Samsung, and Sharp displays, and all 3D home video projectors; passive systems were provided for LG, Toshiba, and some Sony displays, but will not work with home projectors or plasma screen TVs.)

Of course, the diffusion of 3D television depended on the availability of 3D content, of which there was not a lot at the level of broadcast and cable in 2010—and, despite some serious efforts at 3D sports and movie streaming, in the end, very little content materialized. However, affordable 3D Blu-ray discs and players arrived to save the day, as did 3D videogames. The first 3D Blu-ray film title was Sony's *Cloudy with a Chance of Meatballs* in June of 2010. It was followed by *Monster House* in September, *My Bloody Valentine* in October, and in November and December by *The Polar Express*, *Disney's A Christmas Carol*, *The Last Airbender* (M. Night Shyamalan, 2010), *Alice in Wonderland* (Tim Burton, 2010), *Step Up 3D* (Jon M. Chu, 2010), *Despicable Me* (Pierre Coffin and Chris Reynaud, 2010), *Legend of the Guardians: The Owls of Ga'Hoole* (Zack Snyder, 2010), and *Resident Evil: Afterlife* (Paul W. S. Anderson, 2010)—the last three published nearly simultaneously with their 3D theatrical releases. Viewed on large-screen TV's or projectors, the 3D home theater experience could be extremely satisfying, especially with the addition of DTS Master, Dolby 7.1, or eventually Dolby Atmos sound. In fact, the quality and clarity of 3D Blu-rays at home became an issue for RealD theaters where projection could rarely match the discs' brightness and resolution—at least, in the early years of the post-*Avatar* boom.

Postproduction 3D conversion

In the rush to climb aboard the *Avatar* bandwagon, there were a number of hurried, painfully bad 2D to 3D conversions of completed films about to be released in theaters, such as *Clash of the Titans* (Louis Leterrier, 2010), which were nevertheless popular enough to inspire sequels that were also converted to 3D in postproduction, for example, *Wrath of the Titans* (Jonathan Liebesman, 2012) (Figure 6.1). The budgets for these two films—which one critic accurately described as "2.5D movies"—was too excessive even by contemporary blockbuster standards—$125 million and $150, respectively—but they also generated healthy grosses, $798.5 million collectively worldwide in the case of the *Titans* movies, even though their 3D presentations were much less than state-of-the-art. Similarly disappointing from a stereoscopic perspective were *The Green Lantern* (Martin Campbell, 2011), *Captain America: The First Avenger* (Joe Johnston, 2011), and *The Green Hornet* (Michael Gondry, 2011). The first was a Warner Bros. superhero film based on a DC Comics character with a $200 million production budget that barely broke even, ending plans for a sequel (although the character will apparently be revived in the future release *Green Lantern Corps* as part of the DC Extended Universe series).[115] *Captain America: The First Avenger* features a Marvel Comics superhero and is set during the Second World War; produced by Paramount for $140 million, it became a "2.5D" movie when native 3D proved too clumsy and returned $370.6 million worldwide—a disappointment since international marketing costs can consume as much as 40 percent of a film's earnings abroad. *The Green Hornet* was another superhero film, this one from Dynamite Entertainment Comics; produced by Columbia for $120 million, it returned $227.8 million worldwide and made no significant use of 3D technology, which Roger Ebert suggested in his review was "added solely in order to charge extra." The idea that 3D had become a new way for exhibitors to amortize the expense of adding digital projection equipment to their theaters was much in the air in the immediate post-*Avatar* period.

"Native" digital 3D

Yet in January of 2010, James Cameron said that he hoped that *Avatar* would inspire Hollywood to explore the creative possibilities of 3D in genres other than "high-end animation and lowbrow live action," and serious filmmakers were already looking at digital 3D as a way of fulfilling the cinema's promise to completely simulate the real. In early 2011, the three godfathers of American cinema—Martin Scorsese, Steven Spielberg, and Francis Ford Coppola—all announced that they would make their next films in native 3D, as did Ang Lee and Ridley Scott. Such commitments produced Scorsese's *Hugo* (2011), Spielberg's *The Adventures of Tintin: The Secret of the Unicorn* (2011) and *The BFG* (2016), Coppola's *Twixt* (partial 3D, 2011), Ang Lee's *The Life of Pi* (2012) and *Billy Lynn's Long Halftime Walk* (2016), and Ridley Scott's *Prometheus* (2012), *Exodus: Gods and Kings* (2014), *The Martian* (2015), and—as executive producer—*Blade Runner*

115 Six Flags opened two roller-coaster rides called "The Green Lantern" to coincide with the film's release in 2011; both have since been closed.

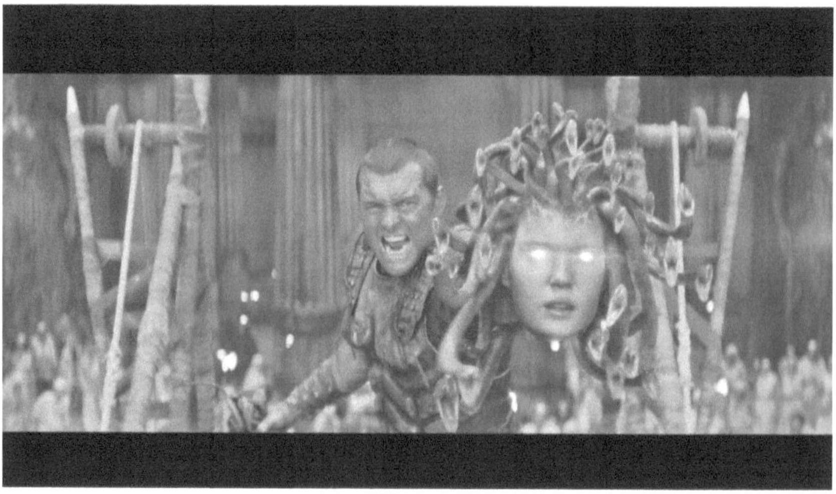

Figure 6.1 *Clash of the Titans* (Warner Bros. Pictures, 2010): shot in 2D with the Arriflex 435 system and converted to 3D (or "2.5 D," as some critics had it) in postproduction for presentation in RealD Cinema. Perseus (Sam Worthington) shows the decapitated head of Medusa (Natalia Vodianova) to ward off the monstrous Kraken.

2049 (directed by Dennis Villeneuve, 2017). Scott's experience in working with 3D was so profound that he said he would never work without it again, "even for small dialogue sequences" (although, in fact, he stopped using it as a director after *The Martian*)—an impression that was borne out by Australian filmmaker Baz Luhrmann's decision to shoot his 2013 adaptation of F. Scott Fitzgerald's *The Great Gatsby* (2013) in 3D (Figures 6.2–6.4). In June of 2009, Peter Jackson announced his intention to shoot all of the films he directed in 3D from then on, which resulted most immediately in the Hobbit trilogy: *An Unexpected Journey* (2012), *The Desolation of Smaug* (2013), and *The Battle of the Five Armies* (2014), and ultimately the brilliantly retrofitted First World War documentary *They Shall Not Grow Old* (2018). Most of these films were carefully produced with a judicious use of digital stereoscopy, but they were surrounded by jerry-rigged mediocrities where 3D, as one critic put it, seemed "little more than a transparent excuse for jacked-up ticket prices."[116]

By mid-2011, a now familiar reaction had set in against 3D. Journalist Mark Harris noted that since *Avatar*, more than half of all 3D releases had

116 Mark Harris, "Honk If You're Sick of 3-D!," *Entertainment*, June 17, 2011.

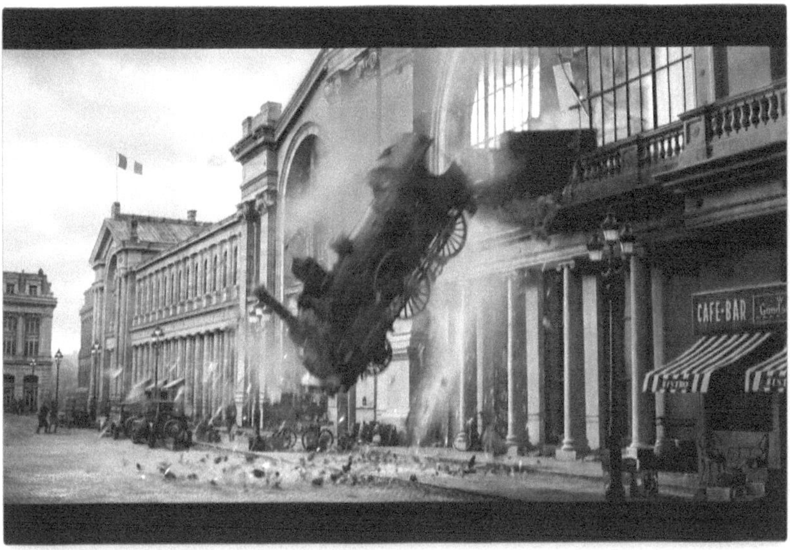

Figure 6.2 *Hugo* (Paramount Pictures, 2011): shot with Arri Alexa Fusion Camera System for presentation in RealD Cinema. A train crashes through the main lobby window of the Gare D'Orsay.

been targeted at kids and teenagers: "For every animated movie that uses 3-D with ingenuity and thoughtfulness [like 2010's *How to Train Your Dragon*] there are three or four that throw an occasional *boi-oi-oing!* at your eyeballs and feel like they've done the job."[117] Worse, the live-action cinema had turned to B-movie material like *Piranha 3D* (Alexander Aja, 2010; a visually satisfying 2D to 3D conversion), *Drive Angry* (Patrick Lussier, 2011; shot in native 3D), *Priest* (Scott Steward, 2011; a sloppy 2D to 3D conversion), and the James Cameron-produced *Sanctum* (Alister Grierson, 2011), an Australian-American survival film that cloned its 3D technology from some of Cameron's previous undersea documentaries; and owing to deceptive advertising practices, it was not always possible for potential spectators to distinguish native 3D from postproduction conversion. Furthermore, there was even some indication at this point that audiences were starting to abandon the 3D issues of films for their 2D versions. For example, the fourth film in the *Shrek* series, Dreamworks' *Shrek Forever After* (Mike Mitchell, 2010), the third in Fox's *The Chronicles of Narnia* series, *The Voyage of the Dawn Treader* (Michael Apted, 2010), and the fourth installment in

117 Ibid.

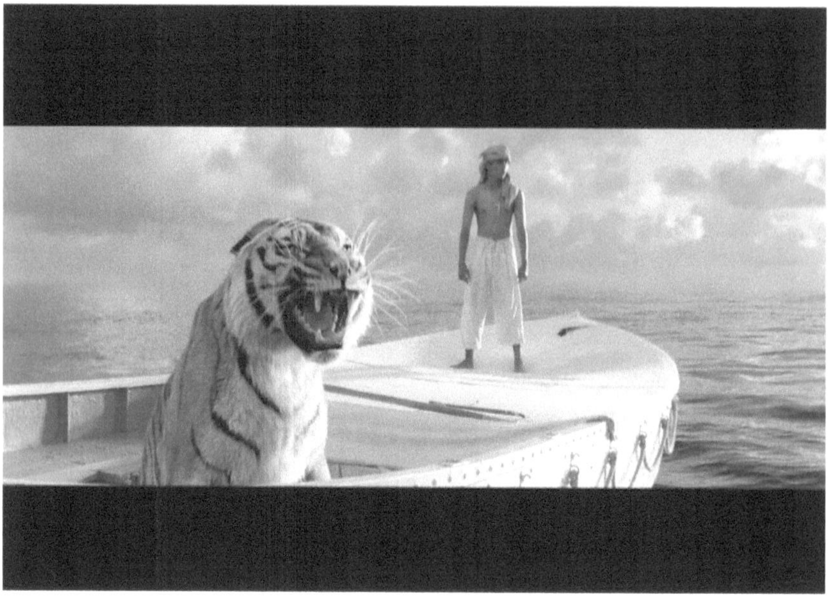

Figure 6.3 *The Life of Pi* (20th Century Fox, 2012): shot with Arri Alexa Fusion Camera System for presentation in RealD Cinema. Pi (Suraj Sharma) at sea with a Bengal tiger in negative parallax.

Disney's *Pirates of the Caribbean* series, *On Stranger Tides* (Rob Marshall, 2011)—despite their inflated 3D admission fees and ultimate profitability—were all the lowest grossers in their respective franchises to date. Part of the problem involved frequent viewer complaints about low light levels in RealD theater projections, and at about the same time some industry analysts began to wonder whether the cost of adding 3D to a production, either by conversion or original design, actually increased its marketability. The rule of thumb in Hollywood was that 3D upped negative costs by about $30 million for an A-list film, and thus the subsequent ticket upcharge for exhibition. But was the upcharge really amortizing the 3D process, if audiences would pay to see the same movie in 2D in roughly the same numbers? As Matt Singer wrote in July 2011, "[I]f only 10% of a film's grosses can be set aside to offset the cost of 3D, a movie earning just 40% of its money from 3D screens would have to be an enormous box office hit to justify its production or conversion in the format."[118]

118 Matt Singer, "3D Movies by the Numbers," July 28, 2011.

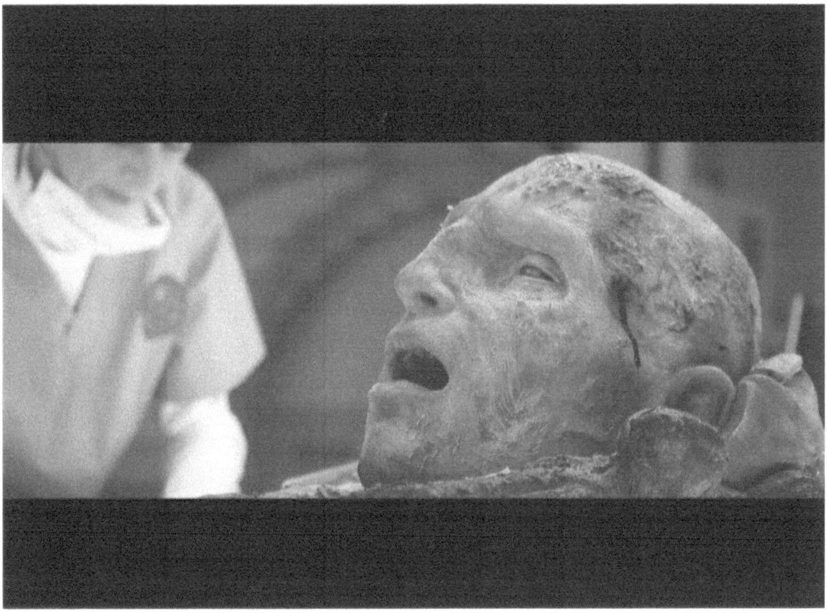

Figure 6.4 *Prometheus* (20th Century Fox, 2012): shot with the 3Ality Technica Atom 3D rig for presentation in RealD Cinema anamorphic 2.39:1 (also blown up to 70mm for IMAX Digital release at 1.90:1). A surgeon (Kate Dickey) prepares to operate on the head of an ancient alien (an "Engineer" of the human race).

Movies from comic books and beyond

In fact, many 3D movies of the second boom were exactly such hits. Some franchises depended on digital 3D because large portions of their audiences expected it. The post-2009 superhero films of the Marvel Cinematic Universe (MCU), the DC Extended Universe (DCEU), and the Marvel-licensed X-Men series—some of which became exponentially profitable—are cases in point. The MCU is a cross-media, shared universe focused on a series of superhero films, independently produced in "Phases" by Marvel and distributed by major studios (most notably Walt Disney), based on characters from Marvel-published comic book, many of them created by the prodigious graphic artist, writer, and publisher Stan Lee (1922–2018). As in the comics, the shared universe was established by crossing over common plots, settings, and casts of characters. The "Phase One" films began production in 2007 with the 2D *Iron Man* (Jon Favreau, 2008) and *The Incredible Hulk* (Louis Leterrier; a reboot of Ang Lee's 2003 version *Hulk*), and continued in 3D with *Iron Man 2* (Jon Favreau, 2010), *Thor* (Kenneth Branagh, 2011), *Captain America: The First Avenger* (as noted above), and the crossover film *The Avengers*

(Joss Whedon, 2012). "Phase Two" consists of *Iron Man 3* (Shane Black, 2013), *Thor: The Dark World* (Alan Taylor, 2013), *Captain America: The Winter Soldier* (Anthony Russo, 2014), *Guardians of the Galaxy* (James Gunn, 2014), *Avengers: Age of Ultron* (Joss Whedon, 2015), and *Ant-Man* (Peyton Reed, 2015). "Phase Three" comprised *Captain America: Civil War* (Anthony and Joe Russo, 2016), *Doctor Strange* (Scott Derrickson, 2016), *Guardians of the Galaxy, Vol. 2* (James Gunn, 2017), *Spider-Man: Homecoming* (Joh Watts, 2017; a reboot of the Sam Raimi's *Spider-Man* trilogy, 2003–2007, and the Marc Webb *Amazing Spider-Man* duology, 2012–14), *Thor: Ragnarok* (Taika Waititi, 2017), *Black Panther* (Ryan Coogler, 2018), *Avengers: Infinity War* (Anthony and Joe Russo, 2018), *Ant-Man and the Wasp* (Peyton Reed, 2018), *Captain Marvel* (Anna Boden, 2019), *Avengers: Endgame* (Anthony and Joe Russo, 2019), *Spider-Man: Far from Home* (Jon Watts, 2019), and the computer-animated *Spider-Man: Into the Spider-Verse* (Bob Persichetti, Peter Ramey, and Rodney Rothman, 2018). These first three phases are collectively known as "The Infinity Saga," and they represent the highest-grossing franchise of all time, having earned $22.5 billion worldwide, with *Avengers: Endgame* displacing *Avatar* as the highest-grossing film in history with global earnings of $2.796 billion. For obvious reasons, the MCU "Phase Four" is currently under strenuous development (e.g., *Black Widow* [Cate Shortland, 2020], *The Eternals* [Chloe Zhao, 2020] were both expected in 2020).

Most of the MCU films were simultaneously released in 2D, 3D, and IMAX 3D. This is also true of the DC Extended Universe series distributed exclusively by Warner Bros. Pictures, which, like the Marvel franchise, also shares common plot elements, settings, casts, and characters. Unlike the MCU, however, the DCEU has a nonlinear crossover story based on the visions of individual filmmakers. Although their worlds are interconnected, the seven films in the franchise so far are distinguished by their individual plot structures. These include *Man of Steel* (Zack Snyder, 2013), *Batman v Superman: Dawn of Justice* (Zack Snyder, 2016), *Suicide Squad* (David Ayer, 2016), *Wonder Woman* (Patty Jenkins, 2017), *Justice League* (Zack Snyder, 2017), *Aquaman* (James Wan, 2018), and *Shazam!* (David F. Sandberg, 2019). The DCEU is the tenth highest-grossing franchise on record, earning $5.28 billion internationally, with *Aquaman* leading the pack at $1.148 billion. As with the MCU, scheduling for further films in the DCEU plows ahead (e.g., *Joker* [Todd Phillips, 2019], *Birds of Prey* [Cathy Yen, 2020], *Wonder Woman 1984* [Patty Jenkins, 2020]). The X-Men superhero film series is based on the group of Marvel comic books by Stan Lee and Jack Kirby, about a group of hyperintelligent mutants, produced and distributed exclusively through 20th Century Fox. There were five 2D X-Men films before the release of the three-dimensional *The Wolverine* (James Mangold, 2013), a highly successful 2D to 3D conversion with a complex range of CGI effects. Two X-Men sequels were produced in native 3D using Arri Alexa rigs—*X-Men: Days of Future Past* (Bryan Singer, 2014) and *X-Men: Apocalypse* (Bryan Singer, 2016), and each was more successful than the last, both commercially and critically. (*The Wolverine* was produced for $120 million and earned $414.9 million worldwide; *Days of Future Past* was produced for an estimated $200 million and returned $748 million worldwide; and *Apocalypse* was made for $178 million, grossing $544 million worldwide.) *Logan* (James Mangold, 2017), the tenth X-Men film and the final installment in the Wolverine trilogy, was not shot or released in 3D, but it did receive a very wide IMAX 2D debut; it became the critical darling of the franchise and the fourth highest-earning R-rated film in history, grossing $619 million against its $127 million budget. The other films of the X-Men cycle—*Deadpool* (Tim Miller, 2016), *Deadpool 2* (David Leitch, 2018), *Dark Phoenix* (Simon Kinberg, 2019), and *The New Mutants* (Josh Boone, 2020)—were released in digital 2D, but with elaborate CGI effects. Collectively, the X-men series represents the seventh highest-grossing franchise in history, with earnings of $6 billion internationally.

These comic-book superhero films whether good, bad, or indifferent had one thing in common—the creation of fantasy worlds—and three-dimensional imagery was uniquely helpful in this regard. In fact, most of the 3D movies released by Hollywood between 2010 and 2020 were in the fantastic genres of horror, extreme action adventure, and science fiction, or simply fantasies per se. As with the 1950s boom, only a handful of "serious" dramatic features appeared in three dimensions, some literary adaptations and some based on original screenplays—for example, *Dolphin Tale* (Charles Martin Smith, 2011), *The Three Musketeers* (Paul W. S. Anderson, 2011), *The Life of Pi* (Ang Lee, 2012), *The Young and Prodigious T. S. Spivet* (Jean-Pierre Jeunet, 2013), *The Great Gatsby* (Baz Luhrmann, 2013), *Gravity* (Alfonso Cuaron, 2013), *The Walk* (Robert Zemeckis, 2015), *In the Heart of the Sea* (Ron Howard, 2015), *Wolf Totem* (Jean-Jacques Annaud, 2015), and *Billy Lynn's Long Halftime Walk Home* (Ang Lee, 2016; 4K, ultra HFR—120 fps) (Figures 6.5 and 6.6). (Some notable 2D to 3D conversions also belong in this category—e.g., the 2012 version of James Cameron's 1997 *Titanic*.)

Of the nearly 600 3D features released worldwide between 2010 and 2020, approximately half were comprised of 3D computer animation or photorealistic CGI, most of them family oriented, while the vast majority live-action films fall into the fantastic genres noted above. The following list will give an indication of their variety (or lack of same).

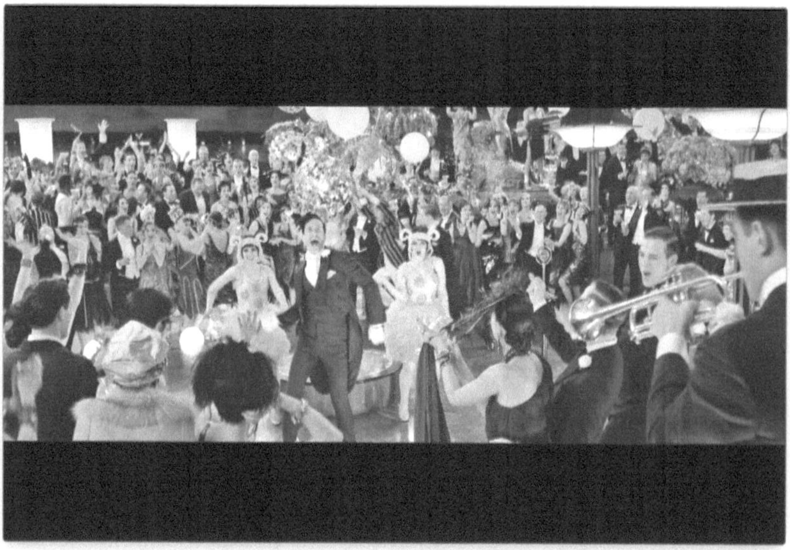

Figure 6.5 *The Great Gatsby* (Warner Bros. Pictures, 2013): shot with the 3Ality Technica TS-5 3D rig for presentation in RealD Cinema anamorphic 2.39:1. A perfectly converged wide-angle shot of a Gatsby party, with the trumpet player at zero parallax.

3D features produced 2010–20, not mentioned as theatrical releases above

Horror

Saw 3D (Kevin Greubert, 2010), *Resident Evil: Afterlife* (Paul W. S. Anderson, 2010), *Haunted—3D* (Vikram Bhatt, 2011; India), *Fright Night 3D* (Craig Gillespie, 2011), *Final Destination 5* (Steven Quale, 2011), *Paranormal Xperience 3D* (Sergi Vizcaino, 2011), *Shark Night 3D* (David R. Ellis, 2011), *Piranha 3DD* (John Gulager, 2012), *Bait 3D* (Kimble Rendall, 2012), *Dracula 3D* (Dario Argento, 2012), *Abraham Lincoln: Vampire Hunter* (Timur Bekmambetov, 2012), *Resident Evil: Retribution* (Paul W. S. Anderson, 2012), *Static* (Todd Levin, 2012), *Underworld: Awakening* (Mans Marland and Bjorn Stein, 2012), *Tormented 3D* (Takashi Shimizu, 2011; Japan), *Sadako 3D* (Tsutomu Hanabusa, 2012; Japan), *World War Z* (Marc Forster, 2013), *Texas Chainsaw 3D* (John Luessenhop, 2013), *Spiders* (Tibor Takacs, 2013), *Nurse 3D* (Doug Aarniokoski, 2014), *Viy* (Oleg Stepchenko, 2014; Russia), *The House That Never Dies* (Raymond Yip, 2014; China), *Dracula Untold* (Gary Shore, 2014), *Poltergeist* (Gil Kenan, 2015), *Paranormal Activity: The Ghost Dimension* (Gregory Plotkin, 2015),

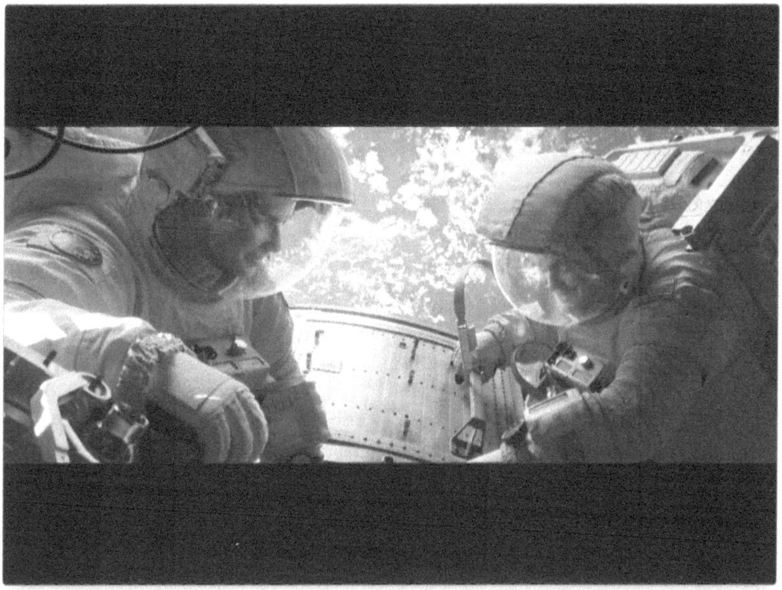

Figure 6.6 *Gravity* (Warner Bros. Pictures, 2013): shot with two different Arri Alexa systems for presentation in RealD Cinema anamorphic 2.39:1. Sandra Bullock and George Clooney floating in space, working on the Hubble Telescope, with the earth behind them.

Underworld: Bloodwars (Anna Foester, 2016), *The Mummy* (Alex Kurtsman, 2017), *Resident Evil: The Final Chapter* (Paul W. S. Anderson, 2017), *The Meg* (Jon Turteltaub, 2018).

Science fiction

Tron: Legacy (Joseph Kosinski, 2010), *John Carter* (Andrew Stanton, 2012), *Upside Down* (Juan Diego Solanas, 2012), *Dredd* (Pete Travis, 2012), *Men in Black 3* (Barry Sonnenfeld, 2012), *Universal Soldier: Day of Reckoning* (John Hyams, 2012), *G. I. Joe: Retaliation* (John M. Chu, 2013), *RoboCop* (Joseph Padilha, 2014; 3D abroad only), *I, Frankenstein* (Stuart Beatties, 2014), *Lucy* (Luc Besson, 2014), *Transcendence* (Wally Pfister, 2014), *Edge of Tomorrow* (aka *Live. Die. Repeat*, Doug Liman, 2014), *Independence Day: Resurgence* (Roland Emmerich, 2016), *Terminator Genesys* (Alan Taylor, 2015), *Jupiter Ascending* (The Wachowskis, 2015), *Passengers* (Morten Tyldum, 2016), *Attraction* (Fyodor Bondarchuk, 2017; Russia), *Blade Runner 2049* (Denis Villeneuve, 2017; also distributed in Oculus VR format), *Valerian and the City of a Thousand Planets* (Luc Besson, 2017), *Ready Player One* (Steven Spielberg,

2018), *The Predator* (Shane Black, 2018), *Venom* (Ruben Fleischer, 2018; an MCU title), *Mortal Engines* (Christian Rivers, 2018), *Goalkeeper of the Galaxy* (Dzhanik Fayziev, 2020; motion capture), *Dune* (Denis Villeneuve, 2020).

Science fiction subgenres and franchises

Disaster films—*The Darkest Hour* (Chris Gorak, 2011), *Sector 7* (Ji-hon Kim, 2011; S. Korea), *Bugs* (Yan Jia, 2014; China), *San Andreas* (Brad Peyton, 2015), *Flight Crew* (Nikolai Lebedev, 2016; Russia), *Geostorm* (Dean Devlin, 2017), *Skyscraper* (Rawson Marshall Thurber, 2018).

Star Trek franchise reboots in 3D—*Star Trek into Darkness* (J. J. Abrams, 2013) and *Star Trek Beyond* (Justin Lin, 2016).

Transformers franchise films in 3D—*Transformers: Dark of the Moon* (Michael Bay, 2011), *Transformers: Age of Extinction* (Michael Bay, 2014), *Transformers: The Last Knight* (Michael Bay, 2017), *Bumble Bee* (Travis Knight, 2018).

Planet of the Apes franchise and reboots in 3D—*Dawn of the Planet of the Apes* (Matt Reevers, 2014), *War for the Planet of the Apes* (Matt Reeves, 2017).

Star Wars franchise and reboots in 3D—the prequel trilogy: *Episode VII—The Force Awakens* (J. J. Abrams, 2015), *Episode VIII—The Last Jedi* (Ryan Johnson, 2017), and *Episode IX—The Rise of Skywalker* (J. J. Abrams, 2019), and the anthology films *Rogue One* (Gareth Edwards, 2016) and *Solo: A Star Wars Story* (Ron Howard, 2018); in 2010, the original six Star Wars films were scheduled for 3D conversion and rerelease in episodic order, but only one appeared—*Star Wars: Episode I—The Phantom Menace* (1999; 3D conversion 2012)—because the others were postponed when Disney acquired Lucasfilm in 2012 and began work on the reboot series note above.

Jurassic Park franchise films—*Jurassic Park 3D* 2013 rerelease, *Jurassic World* (Colin Trevorrow, 2015), *Jurassic World: Fallen Kingdom* (J. A. Bayona, 2018).

Kaiju eiga

Pacific Rim franchise films—*Pacific Rim* (Guillermo del Toro, 2013), *Pacific Rim: Uprising* (Steven S. DeKnight, 2018).

Godzilla franchise films—*Kong: Skull Island* (Jordan Vogt-Roberts, 2017); a prequel to *Godzilla* (Gareth Edwards, 2014); *Godzilla: King of the Monsters* (Michael Dougherty, 2019), *Godzilla vs. Kong* (Adam Wingard, 2020).

Action adventure/combat

Battle of Warsaw 1920 (Jerzy Hoffman, 2011; Poland), *Stalingrad* (Fyodor Bondarchuk, 2013; first Russian IMAX 3D release), *Run 3D* (Simone

Bartesaghi, 2013), *The Crossing* (John Woo, 2014; China/Hong Kong), *Need for Speed* (Scott Waugh, 2014), *Taken 3* (Olivier Megaton, 2014), *Northern Limit Line* (Kim Hak-soon, 2015; South Korea), *Everest* (Baltasar Kormakur, 2015), *Point Break* (Ericson Core, 2015), *Furious 7* (James Wan, 2015), *The Finest Hours* (Craig Gillespie, 2016), *Jason Bourne* (Paul Greengrass, 2016), *The Legend of Tarzan* (David Yates, 2016), *The Fate of the Furious* (aka *Fast & Furious 8*, F. Gary Gray, 2017), *xXx: Return of Xander Cage* (D. J. Caruso, 2017), *The Foreigner* (Martin Campbell, 2017), *Mad Max: Fury Road* (George Miller, 2015), *Operation Red Sea* (Dante Lam, 2018; Chinese), *Mission Impossible—Fallout* (Christopher McQuarrie, 2018), *Gemini Man* (Ang Lee, 2019; high frame rate, or HFR).

ACTION FANTASY (China/Hong Kong, distributed internationally)

Flying Swords of Dragon Gate (Tsui Hark, 2011; remake combining *Dragon Gate Inn* [King Hu, 1966] and *New Dragon Gate Inn* [Raymond Lee, 1992]), *The Taking of Tiger Mountain* (Tsui Hark, 2014), *The Monkey King* (Cheang Pou-soi, 2014), *Dragon Blade* (Daniel Lee, 2015), *Ip Man 3* (Wilson Yip, 2015), *The Monkey King 2* (Cheang Pou-soi, 2016), *The Great Wall* (Zhang Yimou, 2016; most expensive film made in China to date and Zhang's first English-language film); *Crouching Tiger, Hidden Dragon: Sword of Destiny* (Yuen Woo-ping, 2016).

Action comedy

Jackass 3D (Jeff Tremaine, 2010), *A Very Harold & Kumar 3D Christmas* (Todd Strauss-Schulson, 2011), *Spy Kids: All the Time in the World* (Robert Rodriquez, 2011), *Journey 2: The Mysterious Island* (Brad Peyton, 2012; sequel to 2008's *Journey to the Center of the Earth*), *Chinese Zodiac* (Jackie Chan, 2012), *R. I. P. D.* (Robert Schwenke, 2013), *Pixels* (Chris Columbus, 2015), *Skiptrace* (Renny Harlin, 2016), and *Ghostbusters: Answer the Call* (Paul Feig, 2016).

Fantasy

The Last Airbender (M. Night Shyamalan, 2010), *Gulliver's Travels* (Rob Letterman, 2010), *Ghost Rider: Spirit of Vengeance* (Mark Neveldine and Brian Taylor, 2012; based on a Marvel Comics character), *Cirque de Soleil: Worlds Away* (Andrew Adamson, 2012), *Jack the Giant Slayer* (Bryan Singer, 2013), *Oz the Great and Powerful* (Sam Raimi, 2013), *Hansel & Gretel: Witch Hunters* (Tommy Wirkola, 2013), *Percy Jackson: Sea of Monsters* (Thor Freudenthal, 2013), *The Wizard of Oz* (1939; 3D conversion, 2013), *Seventh Son* (Sergei Bodrov, 2014), *Pan* (Joe Wright, 2015), *The Huntsman: Winter's War* (Cedric Nicolas-Troyan, 2016), *Miss Peregrin's Home for Peculiar Children* (Tim Burton, 2016), *Jumanji: Welcome to the Jungle* (Jake Kasdan, 2017), *Into the Rainbow*

(Norman Stone and Gary Mak, 2017), *Pirates of the Caribbean: Dead Men Tell No Tales* (Joachim Ronning, 2017), *A Wrinkle in Time* (Ava DuVernay, 2018), *The House with a Clock on Its Walls* (Eli Roth, 2018), *The Nutcracker and the Four Realms* (Lasse Hallstrom and Joe Johnston, 2018), *Mowgli: Legend of the Jungle* (Any Serkis, 2018), *Jumanji: The Next Level* (Jake Kasdan, 2019), *The Chronicles of Narnia: The Silver Chair* (Joe Johnston, 2020).

Young adult fantasy / science fiction

Harry Potter franchise films—*Harry Potter and the Half-Blood Prince* (David Yates, 2009; partial IMAX 3D); *Harry Potter and the Deathly Hallows—Part 1* (David Yates, 2010), *Harry Potter and the Deathly Hallows—Part 2* (David Yates, 2011), and *Fantastic Beasts and Where to Find Them* (David Yates, 2016)—were all released at feature length in IMAX 3D.

Reboots in the Teenage Mutant Turtles series—*Teenage Mutant Ninja Turtles* (Jonathan Liebesman, 2014) and *Teenage Mutant Ninja Turtles: Out of the Shadows* (Dave Green, 2016)—were both released in IMAX 3D.

The second and third films in The Maze Runner trilogy—*Maze Runner: The Scorch Trials* (Wes Ball, 2015) and *The Maze Runner: The Death Cure* (Wes Ball, 2018)—were both released in large-format 3D (the first one in IMAX 3D).

The second and third films in The Divergent trilogy—*The Divergent Series: Insurgent* (Robert Schwentke, 2015) and *The Divergent Series: Allegiant* (Robert Schwentke, 2016)—were released in large-format 3D (the first one in IMAX 3D).

The third and fourth installments in The Hunger Games series—*The Hunger Games: Mockingjay—Part 1* (Francis Lawrence, 2014) and *The Hunger Games: Mockingjay—Part 2* (Francis Lawrence, 2015)—were both released in 3D.

Videogame franchise films—*Tekken: Blood Vengeance* (Yoichi Mori, 2011; Japan), *Silent Hill: Revelation 3D* (M. J. Bassett, 2012), *Warcraft* (Duncan Jones, 2016), *Assassin's Creed* (Justin Kurzel, 2016), *Tomb Raider* (Roar Uthaug, 2018), *Rampage* (Brad Peyton, 2018), *Pokemon: Detective Pikachu* (Rob Letterman, 2019)—were all released in 3D.

Historicals / mythologicals / biblicals

Conan the Barbarian (Marcus Nispel, 2011), *Immortals* (Tarsem Singh, 2011), *47 Ronin* (Carl Rinsch, 2013), *Pompeii* (Paul W. S. Anderson, 2014), *Noah* (Darren Aronofsky, 2014; 3D abroad only), *300: Rise of an Empire* (Noam Murro, 2014), *Hercules 3D* (Brett Ratner, 2014), *The Legend of Hercules* (Renny Harlin, 2014), *Gods of Egypt* (Alex Proyas, 2016), *Ben-Hur* (Timur Bekmambetov, 2016), *King Arthur: Legend of the Sword* (Guy Ritchie, 2017), *Alpha* (Albert Hughes, 2018).

Crime/mystery

Sin City: A Dame to Kill for (Robert Rodriquez, 2014), *Cold War 2* (Longman Leung Sunny Luk, 2016; Hong Kong), *Murder on the Orient Express* (Kenneth Branagh, 2017; 65 mm).

Concert and dance films

Street Dance 3D (Max Giwas and Dania Pasquini, 2010), *Step Up 3D* (Jon M. Chu, 2010), *The Nutcracker in 3D* (Andrei Konchalovsky, 2010), *Luna Sea 3D in Los Angeles* (Sojiro Otsubo, 2010), *Phish 3D* (Lawrence Jordan, 2010), *Bizet: Carmen in 3D* (Julian Napier, 2011), *Giselle in 3D* (Valery Gergiev, 2011), *Aphrodite Live* (William Baker and Marcus Viner, 2011), *Get Your Sting and Blackout World Tour* (2011), *Glee: The 3D Concert Movie* (Kevin Tancharoen, 2011), Justin Bieber: *Never Say Never* (Jon M. Chu, 2011), *Appetite for Democracy 3D* (Kerry Asmussen, 2012), *Katy Perry: Part of Me* (Dan Cutforth and Jane Lipsitz, 2012), *Satchurated: Live in Montreal* (Francois and Pierre Lamoureux, 2012), *Transcendence* (Bai Quiang, 2012), *Step Up Revolution* (Scott Speer, 2012), *Battle of the Year* (Benson Lee, 2013), *Metallica: Through the Never* (Nimrod Antal, 2013), *One Direction: This Is Us* (Morgan Spurlock, 2013), *Step Up: All In* (Trish Sie, 2014).

European art films

Cave of Forgotten Dreams (Werner Herzog, 2010), *Pina* (Wim Wenders, 2011), *Goodbye to Language* (Jean-Luc Godard, 2014), *Love 3D* (Gaspar Noe, 2015).

Animation (mainly 3D computer)

Walt Disney Pictures/Walt Disney Animation Studios (as producer/coproducer with **Pixar Animation Studios**)—*Tangled* (Nathan Greno and Byron Howard, 2010), *Toy Story 3* (Lee Unkrich, 2010), *Cars 2* (John Lasseter, 2011), *The Lion King* (1994; 3D conversion 2011), *Mars Needs Moms* (Simon Wells, 2011), *Wreck-It Ralph* (Rich Moore, 2012), *Frankenweenie* (Tim Burton, 2012; stop motion), *Finding Nemo* (2003; 3D conversion 2012), *Monsters, Inc.* (2001; 3D conversion 2012), *Beauty and the Beast* (1991; 3D conversion, 2012), *Brave* (Mark Andrews and Brenda Chapman, 2012), *Monsters University* (Dan Scanlon, 2013), *The Little Mermaid* (Ron Clements, 1989; 3D conversion 2013), *Frozen* (Chris Buck and Jennifer Lee, 2013), *Planes* (Klay Hall, 2013), *Planes: Fire & Rescue* (Bobs Gannaway, 2014), *Ratatouille* (Brad Bird, 2007; 3D conversion 2014), *Big Hero 6* (Don Hall and Chris Williams, 2014), *Inside Out* (Pete Docter,

2015), *The Good Dinosaur* (Peter Sohn, 2015), *Finding Dory* (Andrew Stanton, 2016), *Zootopia* (Byron Howard and Rich Moore, 2016), *Moana* (Ron Clements and John Musker, 2016), *Cars 3* (Brian Fee, 2017), *Coco* (Lee Unkrich, 2017), *Incredibles 2* (Brad Bird, 2018), *Ralph Breaks the Internet* (Phil Johnston and Rich Moore, 2018), *Spies in Disguise* (Nick Bruno and Troy Quane, 2019), *Toy Story 4* (Josh Cooley, 2019), *Frozen 2* (Chris Buck and Jennifer Lee, 2020), *Onward* (Dan Scanlon, 2020), *Soul* (Pete Docter, 2020), *Raya and the Last Dragon* (Paul Briggs and Dean Wellins, 2021).

Disney live-action/motion capture, photorealistic CGI remakes of Disney animated classics—*Alice in Wonderland* (Tim Burton, 2010; original was 1951), *Maleficent* (Robert Stromberg, 2014; original was *Sleeping Beauty*, 1959), *Cinderella* (Kenneth Branagh, 2015; original was 1950), *The Jungle Book* (Jon Favreau, 2016; original was 1967), *Alice Through the Looking Glass* (James Bobin; a sequel to *Alice in Wonderland*), *Beauty and the Beast* (Bill Condon, 2017; original was 1991), *Christopher Robin* (Marc Forster, 2018; original was *Winnie the Pooh* films 1997 and 2011), *Dumbo* (Tim Burton, 2019; original was 1941), *Aladdin* (Guy Ritchie, 2019; original was 1992), *The Lion King* (Jon Favreau, 2019; original was 1994), *Lady and the Tramp* (Charlie Bean, 2019; original was 1955), *Maleficent: Mistress of Evil* (Joachim Renning, 2019; sequel to *Maleficent*, 2014), *Mulan* (Niki Caro, 2020; released in streaming version only; original was 1998), Cruella (Craig Gillespie, 2021; released in both theatrical and streaming versions; sequel to 101 Dalmatians, 1961and 1996).

Dreamworks Animation (as producer/coproducer)—*How to Train Your Dragon* (Chris Sanders and Dean DeBlois, 2010), *Megamind* (Tom McGrath, 2010), *Kung Fu Panda 2* (Jennifer Yuh Nelson, 2011), *Puss in Boots* (Chris Miller, 2011), *Madagascar 3: Europe's Most Wanted* (Eric Damell, Conrad Verson, and Tom McGrath, 2012), *Rise of the Guardians* (Peter Ramsey, 2012), *The Croods* (Kirk DeMicco and Chris Sanders, 2013), *Turbo* (David Soren, 2013), *Mr. Peabody & Sherman* (Rob Minkoff, 2014), *How to Train Your Dragon 2* (Dean DeBlois, 2014), *Penguins of Madagascar* (Eric Darnell and Simon J. Smith, 2014), *Home* (Tim Johnson, 2015), *Kung Fu Panda 3* (Jennifer Yuh Nelson and Allessandro Carloni, 2016), *Trolls* (Mike Mitchell and Walt Dorn, 2016), *The Boss Baby* (Tom McGrath, 2017), *Captain Underpants: The First Epic Movie* (David Soren, 2017), *How to Train Your Dragon: The Hidden World* (Dean DeBlois, 2019), *Abominable* (Jill Culton, 2020), *Trolls on Tour* (Walt Dohm and David P. Smith), *The Croods 2* (2020).

20th Century Fox Animation—*Rio* (Carlos Saldanha, 2011), *Ice Age: Continental Drift* (Steve Martino and Michael Thurmeier, 2012), *Epic* (Chris Wedge, 2013), *Rio 2* (Carlos Saldanha, 2014), *The Book of Life* (Jorge R. Guietierez, 2014), *The Peanuts Movie* (Steve Martino, 2015), *Ice Age: Collision Course* (Mike Thurmeier, 2016), *Ferdinand* (Carlos Sadanha, 2017).

Illumination Animation Studio—*Despicable Me* (Mac Guff, 2010), *Hop* (Tim Hill, 2011), *The Lorax* (Chris Renaud, 2012; IMAX 3D), *Despicable Me 2* (Chris Renaud and Pierre Coffin, 2013; IMAX 3D), *Minions* (Pierre Coffin and Kyle Balda, 2015), *The Secret Life of Pets* (Chris Reynaud, 2016), *Sing* (Garth Jennings, 2016), *Despicable Me 2/Despicable Me 3* (Pierre Coffin and Kyle Balda, 2017), *The Grinch* (Scott Mosier and Yarrow Cheney, 2018), *The Secret Life of Pets 2* (Chris Renaud, 2019).

Sony Pictures Animation—*The Smurfs* (Raja Gosnell, 2011), *The Smurfs 2* (Raja Gosnell, 2013), *Hotel Transylvania* (Genndy Tartakovsky, 2012), *Cloudy with a Chance of Meatballs 2* (Cody Cameron and Kris Pearn, 2013), *Hotel Transylvania 2* (Genndy Tartakovsky, 2015), *The Angry Birds Movie* (Clay Kaytis and Fergal Reilly, 2017), *The Emoji Movie* (Tony Leondis, 2017), *Hotel Transylvania 3: Summer Vacation* (Genndy Tartakovsky, 2018), *Wish Dragon* (Chris Appelhans, 2021).

Warner Animation Group—*The Lego Movie* (Phil Lord and Christopher Miller, 2014), *Storks* (Nicholas Stoller and Doug Sweetland, 2016), *The Lego Batman Movie* (Chris McKay, 2017), *The Lego Ninjago Movie* (Charlie Bean, Paul Fisher, and Bob Logan, 2017), *Smallfoot* (Karey Kirkpatrick, 2018), *The Lego Movie 2: The Second Part* (Mike Mitchell, 2019), *The Angry Birds Movie 2* (Thurop Van Orman, 2019), *Scoob!* (Tony Cervone, 2020; released theatrically and streaming).

Paramount Animation—*The SpongeBob Movie: Sponge Out of Water* (Paul Tibbitt, 2015), *Sherlock Gnomes* (John Stevenson, 2018; coproducer Rocket Pictures), *Wonder Park* (Dylan Brown, 2019).

Manga-Inspired 3D CG Anime—*Maya the Bee* (Alexs Stadermann, 2014; Germany/Australia), *Harlock, Space Pirate 3D* (Shinji Aramaki, 2013; Japan), *Ghost in the Shell: The New Movie* (Kazuya Nomura, 2015; Japan), *Dragon Ball Z: Resurrection "F"* (Tadayoshi Yamamuro, 2015; Japan), *Alita: Battle Angel* (Robert Rodriquez, 2019).

Various Other Animation: *Cats and Dogs: The Revenge of Kitty Galore* (Brad Peyton, 2010), *A Turtle's Tale: Sammy's Adventure* (Ben Stassen, 2010), *Alpha and Omega* (Anthony Bell, 2010), *Legend of the Guardians: The Owls of Ga-Hoole* (Zack Synder, 2010), *Yogi Bear 3D* (Eric Brevig, 2010), *Gnomeo & Juliet* (Kelly Asbury, 2011), *Hoodwinked Too! Hood vs. Evil* (Mike Disa, 2011), *The Lion of Judah* (Deryck Broom and Roger Hawkins, 2011), *Arthur Christmas* (Sarah Smith, 2011), *Zambezia* (Wayne Thomley, 2012; South Africa), *Gladiators of Rome* (Iginio Straffi, 2012; Italy), *The Snow Queen* (Vladen Barbe and Maxim Svesnikov, 2012), *Escape from Planet Earth* (Cal Brunker, 2013), *Legends of Oz: Dorothy's Return* (Will Finn and Dan St. Pierre, 2013), *Thunder and the House of Magic* (Jeremy Degruson, Ben Stassen, and Mimi Maynard, 2013), *The Boxtrolls* (Graham Annable and Anthony Stacchi, 2014; stop motion), *Asterix and Obelix: The Mansions of the Gods* (Alexandre

Astier and Louis Clichy, 2014; France), *Yellowbird* (Christian De Vita, 2014), *The Little Prince* (Mark Osborne, 2015; France), *Capture the Flag* (Enrique Gato, 2015; Spain), *Binky Bill the Movie* (Deane Taylor, 2015; Australia), *Mune: Guardian of the Moon* (Alexandre Heboyan and Benoit Philippon, 2015; France), *Robinson Crusoe* (Vincent Kesteloot and Ben Stassen, 2016; Belgium), *Ratchet & Crank* (Kevin Munroe, 2016), *Rock Dog* (Ash Brannon, 2016), *The Son of Bigfoot* (Ben Stassen and Jeremy Degruson, 2017), Asterix: *The Secret of the Magic Potion* (Louis Clichy and Alexandre Astier, 2018; France), *Missing Link* (Chris Butler, 2019; stop motion), *The Addams Family* (Conrad Vernon and Greg Tiernan, 2019).

This list is not comprehensive. For example, it excludes many Chinese/Hong Kong martial arts (*wuxia*), fantasy, and horror films not released outside of Southeast Asia, as well as many Indonesian, Malaysian, Indian, and Mexican features not distributed beyond their domestic markets. The list also excludes 3D documentaries produced specifically for IMAX screens, although more than a few were made during the period in question. Finally, some of the films scheduled for 2020 were not released in 3D as planned because of the COVID-19 pandemic (e.g., Disney's live-action *Mulan* and Warner Bros.'s *Wonder Woman 1984*). Due to theater closures in some regions and restricted access in others, they were released instead to HD video streaming services, where formatting them in 3D would have made no sense given the limited availability of 3D-capable monitors since 2016. Nevertheless, the list contains over 500 titles and suggests the extent of the 2010–20 boom at both the high and low ends of the market.

It is instructive to compare the 1950s "golden age" stereoscopic cinema with the post-*Avatar* boom. The 3D boom (or "boomlet," more accurately) of 1952–55 yielded less than 70 features, and the post-*Avatar* wave of 2010–20 yielded nearly 600 worldwide. Some of these were low-budget and/or one-off productions, but many were blockbuster entries in enormously profitable entertainment franchises, both live-action and computer-animated. Whereas digital technology actually made 3D *less* expensive to achieve as a percentage of total production costs (one camera, one projector, one "film strip" [i.e., one streaming digital feed]), actual negative costs and earnings reached atmospheric proportions, rising into the $100s of millions on each end quite handily. This is because in the second decade of the twenty-first century Hollywood became dominated by branded product.

The American film industry in the twenty-first century had become a crucible for the creation of franchises and brands through the synergistic interaction of corporate parts. A franchise involves the creation of an infinitely exploitable entertainment product, such as the Marvel Cinematic Universe. An entertainment brand has been described as "a lump of content [w]hich can be exploited through film, broadcast and cable television, publishing, theme parks, music, the Internet, and merchandising," like the spokes of a wheel radiating from the branded product so that exploitation of all produces numerous income streams and further strengthens the brand. The effect is to repurpose material created initially for feature film production into other media or to repurpose material created for other media into films. Often, the reproducibility of a film's brand is the key to obtaining production finance. According to film historian Michael Allen, synergy has come to define the modern conglomerate-controlled film industry, so that film subjects are now chosen and developed specifically for their possible connections with other media. As Stephen Prince put it, "Understood in strict economic terms, production by the majors [is] about the manufacture and distribution of commodities (not films) on a national and global scale."

The United States presently has the great advantage of sustaining the largest home market for motion pictures in the world: with more than 40,000 screens, an all-time high, American audiences account for 35 percent of the global box office. Relative to the world market, given that a film's highest costs (those incurred in production) are often amortized in the domestic market, American film revenues abroad can translate into pure profit. Global markets are increasingly important to the transnational entertainment conglomerates that control the US film industry. International distribution has grown from a source of supplemental income to an economic necessity, and risk reduction practiced on a global scale has replaced the guaranteed success offered by the cartelized studio system. In fact, concentration of market power by the American majors has never been stronger in its totalizing nature than it is today, when most national cinemas of the world are dominated by Hollywood films. Among the Western democracies, for example, only countries like France, which has enacted a strong system of subsidies and protection, can hope to control their own markets, which are often swollen with American products by as much as 80 percent.

Market crash

There are numerous signs that in the West, as least, the market appeal of three-dimensional cinema has fallen off. For example, for a short while it seemed as if 3D television had driven the 3D theatrical revolution, but LG and Sony, the last two major TV manufacturers to support 3D features, stopped doing so in 2017, much to the chagrin of committed 3D gamers. Vizio had dropped 3D support in 2013, and Samsung in 2016. DirectTV cancelled its 24/7 3D channel in 2012, and ESPN did the same a year later. In the late 2010s, there was a sense in the television industry that the new 4K UHD and OLED standards would make 3D TV superfluous. (Although, in fact, there is a world of difference between the experience of stereoscopic vs. UHD/OLED viewing on a flat-screen monitor.) Many HD video projectors still are produced with 3D capabilities, and 3D Blu-ray discs continue to be released, although mainly in Southeast Asia, where theatrical 3D is still popular. In fact, 3D does extremely well in some foreign markets, especially China,[119] currently the world's second largest after the United States, and in 2016, 78 percent of the theaters in the Asia-Pacific region were 3D-capable, as opposed to 39 percent in the United States, 43 percent in Latin America, and 45 percent in Europe, the Middle East, and Africa collectively. Some commentators have suggested that Hollywood continues to make 3D films largely due to this enthusiasm for 3D abroad (which accords with the fact that since 2014, some studio movies have only been released in 3D internationally, e.g., *Noah* and *Robo Cop*). At the same time, IMAX announced to its shareholders in 2017 that it would be substantially reducing its 3D offerings in the domestic market, citing a "clear preference" from audiences for 2D.[120] According to multiple sources, fewer than 30 percent of American viewers chose to see the 3D version of a film over its 2D counterpart in 2018, whereas 80 percent of *Avatar*'s worldwide grosses came from 3D in 2009. Obviously, the tables for the popularity of 3D in the United States, at least, have turned.

The eclipse of theatrical 3D was forecast by two monster movies from mid-2019—*Godzilla—King of the Monsters* (Michael Dougherty) and *Crawl* (Alexandre Aja). With a sodden script and less-than-stellar performances, the first was a franchise entry slickly directed by an aficionado of *kaiju* films on a bloated budget of $200 million; it contained state-of-the art 3D effects and sound design. The second was an action-packed *Jaws*-like creature feature involving a post-hurricane alligator attack; it was directed on a budget of $13.5 million by Alexandre Aja, who had shown himself to be a master of low-end three-dimensional terror in the horror comedy *Piranha 3D* in 2010.[121] Released in RealD and IMAX 3D, *Godzilla* earned $386 million worldwide, but it needed to return $550–600 million to break even and thus became a box-office failure. Although eminently suitable to 3D, *Crawl* did not use the process because it didn't need to—its terrifyingly propulsive sequences were effective

119 Jim Van Doren, editor-in-chief and founder of MarketSaw.com: "Chinese audiences love 3D. And, really, they may singlehandedly be enough to warrant the entire industry." (Quoted in Celine Tricart, *3D Filmmaking: Techniques and Best Practices for 3D Filmmakers* (New York: Routledge, 2017), 36.)

120 Carolyn Giardina and Pamela McClintock, "Is the Golden Age of 3D Officially Over?" *Hollywood Reporter*, August 3, 2017. In 2019, however, there were still 1,568 IMAX 3D-capable theater systems worldwide, with the Chinese market bringing in over one-third of the company's earnings.

121 *Piranha 3D* was a remake of Joe Dante's *Piranha*, a grade-B exploitation film produced by Roger Corman in 1978 to cash in on the success of the Jaws franchise; it returned

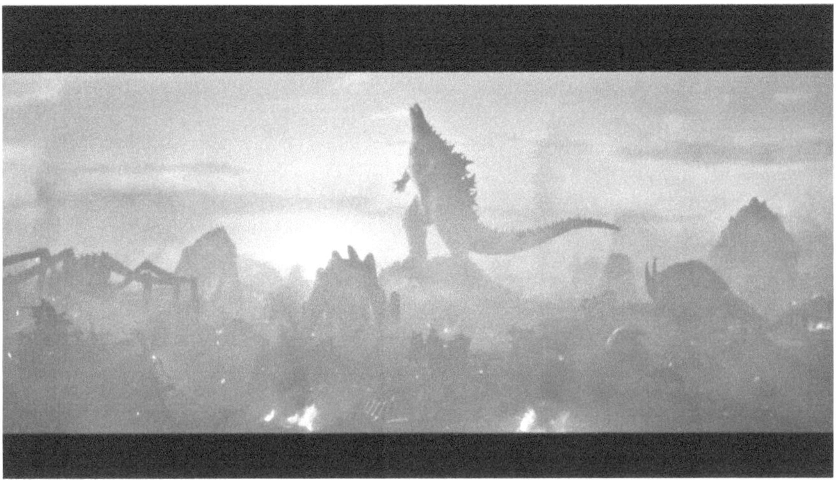

Figure 6.7 *Godzilla—King of the Monsters* (Warner Bros. Pictures, 2019): shot with Arri Alexa 65 and Panavision Primo 70 systems for presentation in RealD Cinema anamorphic 2.39:1. A wide shot in perfect parallax from the conclusion where Godzilla, after epic battles, has conquered Mothra, Rodan, the three-headed King Ghidra, and all of the other "Titans."

enough without it, and the film became a box-office hit when it returned $90 million internationally (Figures 6.7 and 6.8). These facts demonstrated two important things about the state of movie-making c. 2019: (1) big-budget 3D productions could no longer guarantee profits on the basis of spectacular effects alone, and (2) modestly budgeted action films could succeed without recourse to 3D at all.

The falling off in popularity of theatrical 3D in the West happened on many levels. There were the low light levels in auditoria associated with RealD projection lamps and lenses, and there was (as always) the annoyance of wearing glasses—which, whatever other problems they might cause, inevitably put a physical barrier between the viewer's eyes and the screen. There were also the nagging complaints of eyestrain and headaches, never completely resolved by research (see, e.g., Chapter 3). But the main problem seems to

$16 million against a budget of $600,000, and, like its 3D remake, generated a sequel (James Cameron's *Piranha II: The Spawning*, 1982), as well as a TV remake in 1995. *Piranha 3D* was shot in 2D and converted to 3D using a process developed by the company InnerD. Unlike many such conversions, this one was planned by the director in preproduction and well received. The sequel was *Piranha 3DD* (John Gulager), and not nearly as popular.

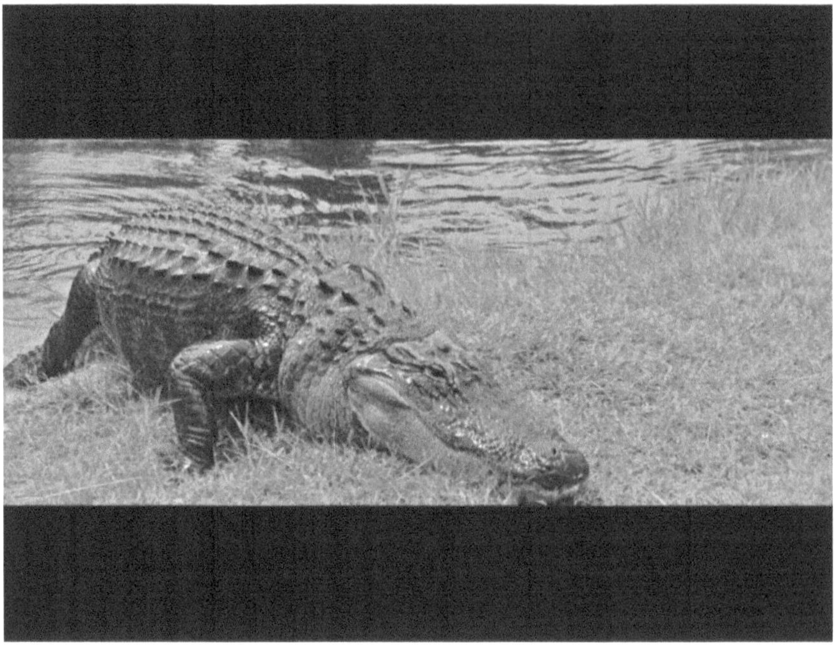

Figure 6.8 *Crawl* (Paramount Pictures, 2019): shot with the Arri Alexa Mini system in 2D and released in 2D at 2.35:1 to RealD theaters. Sometimes giant allegators don't need 3D.

have been the $5-plus per ticket upcharge that theater owners levied on 3D exhibition.[122] This was especially aggravating for the very family audiences at which so many of the fantasy and computer-animated films of the post-*Avatar* era were targeted—not only were kids subject to the upcharge for admission but the one-size-fits-all glasses were not designed for small heads, so that children could not fully appreciate 3D imagery without discomfort.[123] Add to this the 5–10 percent portion of the population who are "stereo-blind"—who cannot achieve stereoscopic convergence for optical (e.g., astigmatism, myopia)

122 There was some justification for the upcharge, including the fact that exhibitors had to purchase large lots of 3D glasses, then manage, distribute, recover, and clean them between screenings.

123 Examples of such films that were successful in 2D but did not perform well in 3D are *Shrek Forever After* (2010), *Kung Fu Panda 2* (2011), *How to Train Your Dragon 2* (2014), and *Maleficent* (2014).

or physical (e.g., vertigo, inner ear dysfunctions),[124] and several demographic segments are practically cut out of the audience for 3D as a regular movie-going experience. There seems to have been some collective wisdom among film producers and TV manufacturers alike near the end of the 2010s that if the public didn't want to pay extra for 3D, maybe they didn't need it at all.

124 According to an article in *In Depth*, July 19, 2012, neuroscientists have found cells in the visual cortex whose function is to transmit the differences in position of the images sent from each eye to the brain. Called "binocular neurons," these cells seem to be the key to seeing in three dimensions. It is thought that the brain may have only a short window opportunity in which to develop these cells, after which it will never form the connections needed to process three-dimensional images and be locked forever into two-dimensional vision (www.bbc.com/future/story20120719-awoken-from-a-2d-world).

Chapter 7

"A DIFFERENT KIND OF MENTAL IMAGE": SOME AESTHETIC CONSIDERATIONS ABOUT 3D

Toward the end of the 2010s, many proponents of 3D thought it had arrived to stay. Some industry insiders were convinced that the process, in its digital form, was the shape of the future, and several scholarly books and articles were written at the time extolling its aesthetic virtues.[125] The most comprehensive

---

125 By the late 2010s there not only was enthusiasm for digital 3D within the industry but a sense that it would eventually replace 2D cinema altogether.

Demetri Portelli, for example, the stereographer for such landmark films as *Hugo* and *The Young and Prodigious T.S. Spivet*, said:

> I believe that just as color film is the "norm" today, all film and television will be shot in digital 3D tomorrow. It just makes sense. The Lumieres [...] wanted color film, they wanted sound, and they wanted depth. All these things now exist and are possible for today's filmmakers. *I believe that 3D is the natural way to experience a film and 2D is often been the compromise.* [my italics]

Jon Landau, the producer of *Titanic* and the *Avatar* franchise, remarked:

> I think the future of 3D is strong; I think it is here to stay. Technology continues to advance both in theaters but also in home and on mobile devices. *Consumers are going to be consuming all of their content in 3D* [my italics]. When you think about it, when you work on a computer, things are already in layers: windows appear in front of another, it's only natural that that experience will be in 3D. 3D is an evolution, not a revolution. Revolution happens overnight, while evolution takes time to happen.

Graham D. Clark, Vice Present and Head of Stereography for Stereo D and Deluxe 3D, observed:

> Right now there's a lot attention on virtual reality, and VR looks better in 3D [...] 3D will move into that arena and then, after that, move into some kind of holographic or augmented reality where that 3D is in your space. Stereoscopy will always be with us, in one form or another. As media become less planar and less monoscopic, 3D will eventually stick and then just naturally be part of any media, and it will be weird not to have stereo in it. We're not there yet; maybe we'll be there in ten or twenty years.

of these was Miriam Ross's *3D Cinema: Optical Illusion and Tactile Cinema* (2015), yet all dealt essentially with two opposing phenomena—positive parallax (images receding from the screen in depth) and negative parallax (images emerging from the screen into audience space).[126] Flat images exist at zero parallax, where objects seem to reside on the screen's surface. While the best stereoscopic scenography attempts to create spatial continuity between objects in positive and negative parallax, it is negative parallax that gets most of the attention. Whatever the degree of continuity, however, the stereoscopic cone of vision is only about 130 degrees. Viewers cannot turn their heads and crane their necks to see what is behind them. (In fact, in most systems significant head movement will destroy the 3D illusion *within* the cone.) The illusion of 360-degree space—and moving through it rather than simply watching it—can only be experienced through the technology of Virtual Reality (VR), which is the next step in the march toward André Bazin's "total cinema" (see Chapter 8).

As Ross suggests, stereoscopy is not merely an effect added to preextant moving images but one that provides "a unique kind of visuality" (a term borrowed from Thomas Elsaesser) by offering an illusory yet simultaneously tangible sense of presence, which is fragile precisely because it is false. Although three-dimensional moving images can be understood as a means of replicating natural human vision, they should also be understood as a form of illusion that makes the stable boundaries of the screen seem permeable and diminishes "the stability of the self in space."[127] This is especially true of IMAX 3D, where the borders of the screen are lost and stretch beyond the audience's field of view in a kind of floating window; in standard digital 3D, the edges of the screen are still highly visible and can conflict with feelings of immersion when an audience is made aware of them.

One of the ways an audience's attention can be called to the borders of the stereoscopic screen is through frame violation. This occurs when an object in negative parallax (emerging from the screen) is cut off or blocked by the edges of the frame, which seems illogical to the viewer's eyes. Ross notes of this phenomenon "the protrusion of objects towards the edge of the frame means that part of the image is cut off and the whole of the object is no longer apparent in negative parallax space," which leads to a flattening of objects and can "produce a disruptive effect as the eyes try to rectify the seemingly incongruous information" that an object is both whole and split across different planes of depth at the same time.[128] Brian Gardner suggests

---

126 Miriam Ross, *3D Cinema: Optical Illusion and Tactile Experience* (London: Palgrave Macmillan, 2015), 2.

127 Ibid., 5.

128 Ibid.

that frame violation "breaks [our] perceptual system—People are not used to having their perceptual systems in that much conflict for long."[129] Furthermore, stereoscopic information takes longer to process than 2D information, so the film editor must take into account the time needed to process stereo images and to adapt our eye to each new shot. In fact, according to Celine Tricart, the respected author of *3D Filmmaking: Techniques and Best Practices for Stereoscopic Filmmakers* (2017), the difference between 2D and 3D editing is the time needed for the brain to analyze a binocular image and restore the associated depth from shot to shot. (This means in general that 3D shots should be longer than 2D shots.)[130]

For this reason, most professionally made 3D films employ a stereographer to design a "depth budget" or a "depth script." This is a plan that plots the depth of scenographic space—that is, the range of negative and positive parallax—for the length of the film and allows the 3D field screen to operate as a continuum.[131] As already established, 3D shooting adds two parameters to traditional film technique: the "interaxial" (IA), or the distance between the two camera lenses, and the angle formed by the optical axis between them, or "convergence." These parameters determine the position of the "stereoscopic window"—a plane positioned at the point of convergence that becomes the future film plane. Everything located between the camera and the window will appear in front of the screen; everything located behind the window will appear behind the screen, dividing it into two areas: negative space and positive space, which together compose the "3D scenic box," a pyramid-shaped cone whose apex is pointed toward the viewer. The percentage of negative relative to positive space constitutes the depth budget—for example, 1 percent negative space and 2 percent positive space yields a depth budget of 3 percent—in other words, a precise measurement of narrative space within the scenic box charted across the entire film. Both overestimating and underestimating the depth budget can cause 3D discomfort—in the first case, eyestrain and headaches; in the second, flat and ineffectual 3D. Depth budgets can vary greatly from shot to shot and are easier to construct for computer-generated animation than live action, where the vagaries of staging are more difficult to control.[132]

Ray Zone has made the analogy between stereographic imaging and musical composition: like a musical composer, a stereographic composer can increase and decrease levels of intensity throughout the film—an effect that can

---

129 Quoted in ibid., 67.

130 Celine Tricart, *3D Filmmaking: Techniques and Best Practices for Stereoscopic Filmmakers* (New York: Routledge, 2017), 193.

131 Ibid., 198.

132 Ibid., 45.

be seen in depth budgets or depth scripts. In their presentation of the variation of objects in positive and negative parallax throughout the runtime of the film, they are like musical scores; and the films themselves resemble audio waveforms that "digitally render the rising and falling volume in an audio piece."[133] By mapping the changes of narrative space in terms of depth and roundness, depth budgets become compositional strategies. Demetri Portelli, the stereo supervisor for Steven Spielberg's *Hugo*, has similarly compared the depth script to a musical score, where "punctuations must be made along the way as to the big moments and the smallest elegant moments; the first layer of this depth script is to ensure that every shot has the 'correct' amount of stereo volume distributed throughout the shot."[134] Speaking on the same theme, Matthew Blute, stereographer for Dmitri Bonderchuk's *Stalingrad* and Michael Bay's *Transformers: Age of Extinction*, sees "the stereographer as a bridge between what is still probably the early days of *volumetric storytelling* [my italics] and the future." Coming from the Greek "stereos" for "solid, firm" and "graphia" for "to write," stereographer is a word that describes "the one who practices the art of representing solids." According to Celine Tricart, in the filmmaking field the stereographer occupies a position similar to that of the VFX supervisor: he/she will be present in preproduction, filming, and postproduction. Setting the mathematical parameters of IA and convergence is only a small part of the stereographer's responsibilities, which should extend from the first reading of the script to the test screening of the final film.[135]

This discourse raises the issue of continuity editing in 3D as opposed to 2D. Working in three dimensions, traditional rules for 180-degree cutting and eye-line matching suddenly change because of the need for a negative–positive parallax continuum across the entire film. As previously noted, IA is the distance between the two camera lenses, real or virtual. The bigger the IA, the deeper the scenic box and the greater the depth budget will be. The IA can range from a few millimeters for a close shot to a foot or more for distant landscapes, and it has to be adjusted depending on the distance of the cameras to the subject in order to conform to the depth budget.[136]

Unlike the IA, convergence doesn't change the size of the 3D scenic box, but it shifts it in and out of relationship to the screen plane or stereo window—the plane where the cameras' optical axes cross one another. The more the convergence angle is increased, the closer to the camera the window will be, but manipulating convergence can cause a geometric distortion in the image

133 Quoted in Ross, *3D Cinema*, 96.
134 Quoted in Tricart, *3D Filmmaking*, 173.
135 Ibid., 181.
136 Ibid., 47.

called "keystoning"—if the fields of vision of two cameras are not roughly parallel, the resulting images are captured with a slight trapezoidal distortion in one eye that is skewed in the opposite direction in the other, a distortion difficult to avoid when shooting with a wide convergence angle.[137] Thus convergence must be constantly monitored both during production and after.

Regarding editing and pacing, many 3D filmmakers believe that when the next generation of viewers gets used to quickly processing stereoscopic images, editors will be able to use faster cuts. At present, however, fast-cutting scenes can prevent audiences from optically fusing 3D imagery. Celine Tricart, for example, observes that the minimum acceptable duration for a shot in 3D is proportional to the similarity of its 3D settings compared with those from the shot that precedes it, but it is rarely less than half a second. The real threshold of stereoscopic vision, she feels, is actually closer to two seconds, because the longer a shot is held, the more time an audience will have to comprehend it. The best 3D editing is often slower than 2D editing, with fewer cuts—for example, Alfonso Curaon's paradigmatic *Gravity* used only 156 shots in its 91 minutes running time.[138] Furthermore, to maintain continuity, the parameters of 3D (depth and roundness) should not change between two matching shots; abrupt changes in the convergence plane can be fixed in postproduction through "depth balancing," in which the convergence plane of shot number 1 is gradually moved closer to the position of shot number 2, by horizontally shifting the left and right frames (a process known as "horizontal image translation"). Such dynamic shifts of convergence are usually done in the laboratory during the so-called "3D PASS,"[139] which also includes geometric correction, color matching, and color grading, followed by the addition of credits, subtitles, and any other printed matter.[140]

Focal length is another parameter that strongly affects 3D in terms of the roundness of objects within the scenic box. Depth-of-field ironically represents a problem for 3D: depth is necessary to lend objects dimensionality, but too much depth-of-field and too much detail in the background causes the eye to wander all over the screen.[141] (It can also cause problems with tracking shots through movement parallax, especially in negative space.) The longer the lens, the flatter the shapes will be, and beyond a certain distance the elements of the frame will appear as cardboard cutouts. The impact of long lenses on roundness is caused by the apparent shortening of the distance from the

---

137 Ibid., 52.
138 Ibid., 168.
139 Ibid.
140 Ibid., 183.
141 Ross, *3D Cinema*, 29.

camera to the subject, so that the depth budget becomes a necessary tool to produce roundness in terms of volume and texture.[142]

Thus, the choices of IA, convergence, focal length, and roundness must be consistent and serve the narrative. As Richard W. Kroon, CCO for the 2D/3D conversion company Southbay, notes:

> It takes an audience member a certain amount of time to understand and fully appreciate a new visual image following a cut [...] Attention should be paid to the depth characteristics of adjoining shots so that the audience does not experience an unintended jarring change. In 2D we read images more easily, however complex they might be. We have an overdeveloped [screen] culture of 2D imagery.[143]

This is because in 2D filmmaking, over time, film artists became masters of 3D to 2D conversion, that is, they lived and worked empirically in a three-dimensional world that had to be converted to a flat screen, relying on traditional two-dimensional depth cues; for stereoscopic cinema, the process has to be enhanced and modified.[144]

Addressing the matter of stereoscopic errors Miriam Ross writes that misalignments and contradictory spatial cues can break apart the continuity of depth and the coherence of space in the image. She notes, for example, that objects in negative parallax will sometimes bulge (i.e., keep their spatial relationship with their zero- or positive-parallax components, like the wall drawings in Werner Herzog's *Cave of Dreams*),[145] because they retain their frontality. In fact, VR notwithstanding, our visual experience is in film always frontal, because we are always in the position of encountering a body that is opposite us. For this reason, many scholars of 3D take the screen to represent the fourth wall and identify the power to push objects into negative parallax as a way of breaking the screen—as "the cinematic frame being pierced."[146] This applies most obviously to the case of dispersed "stereoscopic debris"—a fragmented mass of material, such as dust, snow, or other particulate matter, that moves from a frontal plane of action toward the audience to produce what Stephen Bottemore calls a "looming" response (flinching and ducking).[147]

142 Chris Parks, stereographer for *Jack the Giant Slayer*, *Gravity*, and *Pan*, quoted in Tricart, *3D Filmmaking*, 59.

143 Quoted in Tricart, *3D Filmmaking*, 193.

144 Cory Turner, Head of Paramount Preproduction, quoted in Tricart, *3D Filmmaking*, 139.

145 Ibid., 61.

146 Ibid., 63.

147 Cited in Ross, *3D Cinema*, 131.

Within 3D films, "the presence of elements in negative parallax space offers an immediate sense of excess, as their sensory engagement present the possibility of an embodied relationship between film and viewer that is surplus to narrative cognition"[148] This notion of "stereoscopy's sensory excess" comports with the fact that it has historically been bundled with other film technologies that enhance experiential engagement—for example, with the Polyvision triptych process in the 1920s (see Chapter 2) and with Cinerama, Cinemascope, and VistaVision in the 1950s (see Chapter 3). Finally, Ross writes, "Stereoscopic sight means confusing the eye and tricking the brain, and [...] is a cognitive effect rather than a visual one."[149]

Aesthetics and some technical matters

Projection

Before digital technology, 3D projection required two perfectly aligned and synchronized projectors.[150] The digital age led to a resurrection of 3D due to the prospect of projecting 3D with a single projector, but there are three flaws inherent in digital projection that can still interfere with visual comfort and the overall 3D experience.[151] (1) **Flicker**, the judder or strobing effect particularly visible during traveling and panning shots. The movement appears to be intermittent or flickering due to the projector's frame rate of 24 fps, each frame being projected two or three times and alternating between the left and right eye. Such "motion blur" results when the viewer's brain confuses the retinal disparity inherent in 3D with the horizontal displacement of the corresponding object on the retina, causing the flicker effect. To solve this problem the shooting and projection rates must be higher than 24 fps, either 48 or 60, such as the 48 fps Peter Jackson used for *The Hobbit Trilogy*. The downside of this High Frame Rate (HFR) is a "hyper-realistic" effect that can make the film image appear like a live TV broadcast or a computer animated cartoon."[152] (2) **Ghosting**, due to the fact that 3D glasses do not entirely block the light from one eye to the other. High contrast between objects with strong parallax may pass through both lenses, causing a "double image." The amount of ghosting depends not only on the projection system but also on the overall contrast and depth budget of the film. (3) **Lack of brightness**: all

148 Ibid., 59.
149 Ibid.. 41.
150 Tricart, *3D Filmmaking*, 93.
151 Ibid., 100.
152 Ibid., 101–-102.

2D projections have to comply with a strict minimum luminance standard of 14-foot lambert (ft-L) per screen, but as of 2017 the Society of Motion Picture and Television Engineers (SMPTE) had set no official standard for the minimum luminance for 3D screens. 3D digital projection filters create unsatisfyingly low light levels in theaters that do not compensate for the loss by boosting their lamps' brightness (an expensive proposition for most exhibitors). As of 2017, most non-IMAX 3D screens were measured at 3-ft-L or less, or four times darker than 2D. To deliver more lumens to the screen, some theaters—notably IMAX-certified ones—began installing 3D laser projectors in 2012 that deliver an impression of depth and sharpness to the image ten times that of standard digital 3D.[153]

Digital 3D distribution standards and formats

Before the development of 4K (and, soon, 8K) standards, few formats allowed a full HD resolution for each eye, since most "compress" the left and right image together either horizontally or vertically into a single digital feed, because most distribution pipelines were capped at the then-standard HD resolution of 1920 × 1080 pixels. (Since that time, bandwidth and workflows have been upgraded in most cases to 4K resolution.) This compression could take the form of side-by-side, over-under, or the now rarely utilized interlaced files for each frame. Another digital 3D format is Blu-ray disc,[154] which encodes the "master" eye at full resolution and extrapolates the "slave" eye information to produce full HD resolution for both eyes. Since 3D is mainly a matter of horizontal parallaxes, coding the differences between the left and right eye in the disc transfer process conserved bandwidth but also allowed for high-quality stereoscopic reproduction.[155] However, the viability of this format may soon be in question—at this writing, Samsung, a major manufacturer of Blu-ray players, had announced that it will stop making them in 2020,[156] and high-definition streaming offers serious competition for disc playback at the visual level, if not yet in audio terms.

Theatrical 3D distribution is now carried on through DCP, which stands for "Digital Cinema Package" and is the only international standard for streaming digital cinema to theaters (or, initially, downloading discs to projectors). A regular 2D DCP consists of a JPEG2000 still for each frame,

153 Ibid., 102–105.

154 Of course, Blu-ray is also a high-definition 2D format, vastly superior to DVD, which it has virtually replaced at the consumer level.

155 Ibid., 108.

156 *The Wall Street Journal*, October 12–13, 2019: D10.

and 3D DCP is exactly the same except that it alternates a left and right frame in each image sequence. Cinema servers then project the frames alternately, the same left–right pair being projected twice in the case of a "double-flash" projection, or three times in a "triple-flash" projection. DCP 3D is currently (*c.*2018) the best existing 3D format, with no resolution loss and a very low compression ratio for the individual files.[157]

2D to 3D conversion

Conversion is a reasonable alternative when shooting native 3D with a multiple-camera rig that is too expensive or technically challenging (e.g., for aerial shots or for tight spaces like the inside of a car). Sometimes native and conversion are used in the same film—a couple of converted shots within a native 3D scene will often go unnoticed, which is known as "hybrid 3D."[158] The basic idea is to extrapolate a second camera position from the 2D version of a film for a "second eye." In the best conversion practice, the original 2D shot goes through a virtual reconstruction of the space via software like Maya or 3DSMAX. The original textures are then applied to the reconstructed volumes, and the missing parts filled in by hand. Then, the shot is "refilmed" with the use of a virtual 3D rig and the assistance of a stereographer for creative control. A less complicated and expensive conversion technique is the creation of a "depth map"—a grayscale version of each shot, where each gray value corresponds to a given horizontal offset of the associated pixel, thus placing it in depth (e.g., Disney's animated *The Lion King* was converted in this way). The amount of automation and the quality of the algorithms employed will determine the final quality of the conversion, which will ideally be as flawless as native 3D. Conversion has other advantages—for example, it shifts all stereoscopic decisions to postproduction and can take into account traditional 2D editing practices, which need to be modified for 3D as noted above—although converted works can sometimes distort screen geometry. However, production for 2D to 3D conversion means shooting without any 3D monitors on set, whereas shooting in native 3D allows for the optimal creative use of stereoscopy and permits the filmmaker to see the 3D results on set right away.[159]

In both aesthetic and practical terms, three-dimensional cinema is fundamentally different from two-dimensional film. It requires different styles of editing and camera movement, and, fundamentally, it demands the careful

157 Tricart, *3D Filmmaking*, 108.
158 Ibid., 85.
159 Ibid., 86.

management of the depth perspective across the entire range of a work via a professionally created depth budget or depth script. The unique vectors of interaxial distance, convergence, and the maintenance of the parallax continuum in combination with traditional depth cues make the circumstances of 3D production less addenda to 2D production than something sui generis. Jennifer M. Barker suggests that 3D audiences, like those in flat cinema, are "doubly situated," both immersed in the film on-screen *and* aware of the theater space that surrounds them, so that viewing becomes a process of oscillation between these two states.[160] Yet Miriam Ross adds that "there is an additional doubled cognitive function at work in stereoscopic viewing," which she calls "haptic visuality [...] whereby textures and surfaces are made tangible to the senses even though they reside within optical illusions."[161] Both theoretically and practically, then, three-dimensional cinema is a very different beast than flat film. The 3D filmmaker must constantly be aware of the "instability" of stereoscopy's optical illusion in creating three-dimensional imagery and respect that imagery's uncanny nature in the process of manipulating it.

160 From Jennifer M. Barker, *The Tactile Eye* (2009), quoted in Ross, *3D Cinema*, 43.
161 Ross, *3D Cinema*, 45–46.

Chapter 8

"EXPERIENCE ON DEMAND": VIRTUAL REALITY

While digital stereoscopy represents the most recent manifestation of a process that began with Charles Wheatstone's stereoscopic line drawings nearly two centuries ago, it is not the end of it. The ultimate technology for seeing things in three dimensions is Virtual Reality (VR), which uses a hybrid of advanced modern technology—Lidar scanners, hyper-accelerated graphic cards, and so on—and the stereoscopic illusion first quantified by Wheatstone to create a state of sensory immersion that borders on otherness. Finding a way to mass-market the VR experience as a form of popular cinema, rather than as an enhanced form of video game, has become the new Grail Quest of the film industry. But VR also has a more elevated, philosophical justification.

The iconic French film theorist André Bazin believed that each new technological development in the history of cinema (e.g., the introduction of sound, color, widescreen, and 3-D) brought it closer to its ultimate destiny—the production of a total simulacrum of realty—which was implicit in its very beginnings. As he wrote in 1946:

> The idea of Cinema tends toward a full and total representation of reality, and immediately considers the dream of a perfect illusion of the outside world with sound, color, and depth. If the cinema in its infancy did not have all the attributes of the total cinema of tomorrow, this was no fault of its own, and was only because if fairies were technically impotent and couldn't give it to them even though they wanted to.
>
> The guiding myth, then, inspiring the invention of cinema, is the accomplishment of that which dominated in a more or less vague fashion all the techniques of the mechanical reproduction of reality in the nineteenth century, from photography to the phonograph, namely an integral realism, a recreation of the world in its own image, an image unburdened by the freedom of interpretation of the artist or the irreversibility of time. [...] Every new development added to the cinema

> must, paradoxically, take it nearer and nearer to its origins. In short, cinema has not yet been invented! ("The Myth of Total Cinema," in André Bazin, *What Is Cinema?*, vol. 2, trans. Hugh Gray [University of California Press, 1967])

Reviews of Alejandro G. Inarritu's six-and-half minute VR installation at the 2017 Cannes Film Festival seem both to echo and answer Bazin's declaration. *Carne y Arena* (*Flesh and Sand*) places the spectator ("participant" might be a better term) in the desolate predawn wilderness of the Sonoran Desert as a member of a ragged group of immigrants attempting to cross the border into the United States. Suddenly a helicopter appears in the sky, shining its spotlight downward, and as it flies closer and closer, the light becomes blinding, the sound overwhelming; soon, border agents with guns and dogs are shouting at people to put their hands up in two languages. Then, as one reviewer put it:

> You find yourself doing something that's utterly instinctual: You duck. You try to get out of the way. The threat is that physical, that scary, that real. Just like that, you've become part of the audience cowering to avoid being shot at in "The Great Train Robbery." [The author refers to the final shot of a 1903 one-reeler by Edwin S. Porter in which an outlaw fires his revolver directly into the camera lens, which caused contemporary audiences to panic.] Giving yourself over to Inarritu's technically extraordinary, surprisingly humane experiment is an event so novel that it's a little like being at a nickelodeon 100 years ago. (Owen Gleiberman, "Cannes Virtual Reality Review: Alejandro G. Inarritu's 'Carne y Arena'," *Variety*, May 20, 2017)

This reprises Bazin's notion that "Every new development added to the cinema must [...] take it nearer and nearer to its origins." Yet another reviewer of the installation concluded that "virtual reality [is] a medium that seeks to become a non-medium—a tissue of images and sounds that replicates or even supersedes true life" (Jason Farago, "Inarritu's 'Carne y Arena': Virtual Reality Simulates a Harrowing Border Trek," *Variety*, May 19, 2017).[162] In

162 The experience of *Carne y Arena* begins in a cold room strewn with dilapidated shoes. When an alarm sounds, viewers proceed one by one into a large room, put on HDMs, and for six-and-a-half minutes wander barefoot on real sand across a darkened virtual desert populated by scattered migrants who run for cover when a helicopter swoops down and gun-bearing Border Patrol agents begin taking prisoners. Later, in a long hallway, the viewer sees portraits of the immigrants from the VR room, along with their personal histories. The installation has since toured in cities in the United States (e.g., Los Angeles and Washington, DC) and won a Special Achievement Award from

other words, the "recreation of the world in its own image" Bazin saw as the beginning and end-point of film history. Does VR, then, as the ultimate form of stereoscopy, foretell the end of narrative cinema? Or does it look forward to the birth of the "total cinema" that Bazin in 1946 declared had not yet been invented?

Despite the notable success of *Carne y Arena*, at this writing VR has not yet become a sustainable medium of mass entertainment, and questions remain about the effect VR will have on the film industry as a whole. Will it integrate naturally with the existing media of movies and television, or will it bloom into a truly revolutionary medium of its own that opens up whole new markets and infringes the market shares of others? Shortly before the appearance of Inarittu's installation at Cannes, Arthur Van Hoff, chief technical officer, made this important observation:

> Every time there was a disruptive technology shift in movies, there were a lot of naysayers. A lot of people didn't believe in talkies, or in color, or in digital. The same thing is true with virtual reality, which is just another step in making the movie more immersive. I don't think that virtual reality is going to replace filmmaking tomorrow, but in five to ten years, it will have a huge effect on the way movies are made.[163]

In 2021, most VR experiences require the user to wear a head-mounted display (HMD) that positions a pair of specially calibrated stereoscopic images in front of his/her eyes while integrated headphones deliver binaural, three-dimensional surround sound. Motion sensors track head and body movement, adjusting visual and sonic inputs accordingly, so that the user has the illusion of being fully immersed in a 360-degree alternate universe—a virtually real one.

Histories of VR trace its origins back to Wheatstone and Brewster's mid-nineteenth-century experiments with stereography and proceed through the subsequent development of stereophotography and the ubiquity of the Holmes stereoscope in middle-class households during the 1870s. Special mention is made of the grand panoramas and cycloramas of the late nineteenth century as particular kinds of immersive entertainments, especially Grimoin-Sansons's Cinéorama, devised for the Paris Exposition of 1900, which combined a

---

the American Academy of Motion Picture Arts and Science, the first in over 20 years. (The previous one had gone to John Lassiter for directing *Toy Story*, the first feature-length computer-animated film, in 1995.) The Academy has only given 16 such awards in its 94-year history.

163 Quoted in Graham Edwards, "The Dreamsmiths Unleashed," *Cinefex*, 151 (February 2017): 98.

rotunda with 360-degree cinema projections to simulate a ride in a hot-air balloon over the city (see Chapter 1). Abel Gance's multicamera/projector Polyvision process is also mentioned as a prototype (see Chapter 2), as is Fred Waller's Flexible Gunnery Trainer, which used a spherical screen and five projectors to simulate bomber flight during the Second World War, complete with realistic sound from headphones and vibration feedback through dummy guns for trainees (see Chapter 2). The wraparound Cinerama tryptic screen of the early 1950s, with its seven-track stereophonic surround sound, is also invoked as a VR model, and, of course, the 3D theatrical film experience from 1952 through the present is another, very prominent one (see Chapter 3). VR, however, is not at present a collective experience but an individual one, so another model would be the View-Master stereoscopic slide viewer, a twentieth-century redesign of the Victorian stereoscope that has remained in production for almost 90 years as both a popular consumer entertainment toy and educational tool (see Chapter 2).

The first person to use the term "virtual reality" was actually Morton Heilig (1926–97), inventor of a head-mounted cathode-ray tube (CRT) device called the Telesphere. Patented in 1960, this instrument provided 140-degree stereoscopic views of moving images by positioning two miniature televisions in front of the eyes; stereophonic sound was carried by a built-in headphone, and a pair of air discharge nozzles could simulate spatial displacement, changing temperatures and scents. In 1962 Heilig introduced the Sensorama, an immersive multisensory machine with stereoscopic color displays, multiple speakers, air valves to simulate wind and odors, and a vibrating seat for tactility. The viewer would sit in a motional chair and watch one of five films, four of which depicted first-person rides (by bicycle, motorcycle, go-kart, and helicopter) and one of which simulated interaction with a human subject (a belly dancer, prefiguring VR porn). Heilig hoped to sell his "Experience Theater" to industry for the purpose of training and education along the lines of Waller's flight simulator, but it never became commercially viable.

In 1968, Ivan Edward Sutherland (b. 1938) created the first HMD that could track a viewer's gaze in "The Sword of Damocles," so called because the helmet was so heavy that it had to be suspended from the ceiling by a mechanical arm. Although its stereoscopic images were simple wireframes, the HMD had a see-through feature of the sort used by US military helicopter pilots so that the viewer could locate their position in the room. Sutherland, who had invented the first graphical user interface in the Sketchpad program in 1962, went on to form, with David C. Evans, the Evans and Sutherland computer graphic company, which became one of the largest distributors for dome-shaped spherical immersive presentations in the United States, employing both the future founders of Adobe (John Warnock) and Silicon Graphics (Jim Clark).

At the beginning of the 1980s, the history of VR intersected with that of interactive video games. In 1982, the Vectrex system used a one-inch monochrome CRT capable of displaying vector stereographics when used in conjunction with 3D goggles; it featured one built-in game but could also accept playable cartridges, and it was the first such game that did not need to be hooked up to a television set.[164] In 1991, Sega announced a VR HMD that would contain inertial sensors and active-shutter LCD screens in its visor, as well as stereo headphones; it was intended to accompany its Genesis game console but was not released because its prototype reportedly caused motion sickness and headaches in users. In 1995, Nintendo introduced Virtual Boy, a 32-bit table-top game console that displayed monochrome stereoscopic images in parallax when viewers placed their heads against an LED eyepiece; an expensive commercial failure, Nintendo abandoned the console in 1996, although it became the model for the company's 3D-capable GameCube in 2011 and its autostereoscopic, handheld Nintendo 3DS in 2011.

VR entered the arcade in 1991 with Virtual Group's "Dactyl Nightmare," a real-time multiplayer game that employed stereoscopic visors with four-speaker headphones and a microphone, joysticks, and waist-high belts with receivers for interaction with other players and with the computer controls. The graphics, which represented attacking pterodactyls and defending grenade-launchers, were basic polygons, but the game's tracking capacity was significant, and it generated several others that included the popular first-person shooter "Grid Busters." Sega's experiment with console-based VR gaming produced a viable HMD that found its way into several SegaWorld arcade attractions, most notably "Sega VR-1" and "Virtual Racing," both of which featured sensors for tracking head movement and stereoscopic 3D polygon graphics.

At about the same time, VR began turning up in theme parks and similar entertainment venues. One of the first was "Aladdin's Magic Carpet Ride," which was installed at EPCOT in Orlando, Florida, in 1994 and opened with an improved HMD at DisneyQuest interactive theme parks in several US cities. Even earlier, Disney had introduced a process called Circle-Vision 360 in Circarama theaters in a number of its other theme parks. Similar to Grimoin-Sansons's 1900 Cinéorama, Circle-Vision uses nine screens arranged in a

164 Originally called the "Mini Arcade," the Vectrex system tried to match the arcade experience by providing colored plastic overlays with each game, some of which were quite good; these included the factory-installed "Minestorm," plus "Web Wars," "Polar Rescue," and "Cosmic Chasm."

circle around the viewers and nine projectors at head level placed in the gaps between them to cast their images across audience space onto the screens; titles in various versions included *America the Beautiful* (1955–79), *Wonders of China* (1982–2003), *American Journeys* (1984–94), and *The Timekeepers* (1994–2006). In other attractions, Disney continues to use stereoscopic film and motion simulation to create immersive experiences, such as "Honey I Shrunk the Audience," which seems to reduce the scale of the theater, and "Soarin' over California," which involves the audience in simulated flight inside of a giant stereoscopic dome theater. Universal Studios theme parks also offered immersive, VR-like experiences, for example, "T2 3D: Battle across Time" (1996–2017), directed by James Cameron at a cost of $24 million (the production budget for the 12-minute film; the whole attraction, including a specialized theater, cost $60 million). Universal's "Back to the Future" ride functioned as a kind of VR mini-sequel to Robert Zemeckis's 1990 theatrical feature *Back to the Future, Part III.* With costs similar to "T2 3D," the ride suspended audiences in front of an 85-foot IMAX dome and was produced by Steven Spielberg; it ran from 1991 through 2007, when it was replaced by "The Simpsons Ride" in the same venues. In 2017, Samsung teamed with Oculus to produce a VR rollercoaster ride for Six Flags amusement parks with riders wearing HMDs in moving rollercoaster cars.

An extension of these ideas was the cave virtual environment, known as CAVE and developed at the University of Illinois in 1991, where the walls of a room were made up of rear-projection video screens (and now flat panel screens) containing real-time 3D graphics. Motion capture was used to record the position of the viewer, who wore active-shutter LCD glasses to see computer-generated 3D images that seem to float before him/her in tandem with stereophonic surround sound. Sensors allowed the glasses to be tracked, so that the projected images matched the viewer's moving point of view and created the illusion of standing in a real environment. In a further elaboration, the StarCAVE at the University of California, San Diego, uses 17 screens to create a five-walled space, with each wall three screens high and two screens for the floor. High-resolution 3D imagery is created by 34 projectors, and the wall screens tilt toward the viewer progressively at 15-degree angles, so that there is the illusion of complete immersion in 360-degree space. CAVEs have been used extensively for the purpose of industrial design and construction engineering.

In 1977, the first data glove, called the Sayre Glove after its designer Rich Sayre, was created by Electronic Visualization Laboratory (EVL) at the University of Illinois at Chicago. This was a wired glove used as an input device for human–computer interaction—it allowed hand gestures to control data by sensing finger flexion. In 1982, Thomas Zimmerman at VPL Research patented an optical flex sensor that could be mounted in a glove to measure finger bending, and this led to VPL's development of the DataSuit, the first full-body tracking system for VR. VPL was the first company to market head-tracking HMDs and to license data gloves to video game manufacturers (first as the DataGlove to Mattel and then the PowerGlove to Nintendo, both in 1989).

Other research companies were developing VR products for scientific and military applications, among them Fakespace Labs founded in 1991 by Ian McDowall, Mark Bolas, and Eric Lorimer. Fakespace contributed such innovations as the PINCH Glove System, which allows users to "grab" virtual objects or initiate interaction with a finger snap, and the BOOM (Binocular Omni-Orientation Monitor), which located a stereoscopic computer monitor at the end of a boom arm from which a user could view high-resolution imagery through six axes of motion. In 2006, McDowall collaborated in the development of a wide-angle stereoscopic HMD called Wide5 that weighed less than 2 lbs and commanded a 150-degree field of vision; other projects included high-frame-rate projectors and light field displays for VR imaging.

The Wide5 cost about $32,500 per headset, but much less expensive HMDs were in the offing. At the low end, this took the form of the FOV2GO, a "do-it-yourself" handheld VR viewer developed in 2012 at the University of Southern California that harnesses the computer graphics power of smartphones. Instructions for constructing one at home are available on the USC Institute for Creative Technologies website (www.projects.ict.usc.edu/mxr/diy/fov2go-viewer), but roughly one million *New York Times* subscribers received their own Cardboard VR viewer with their November 8, 2015, Sunday edition. Based on Google's open-source design, this version of the FOV2GO was a paper-based headset compatible with any Android or iOS phone, and it introduced the new app NYTVR with "The Displaced," a VR documentary short that took viewers inside the lives of three children displaced in the global refugee crisis; it was followed by an immersive dispatch from the streets of Paris in the aftermath of the November 13 terrorist attack and the fanciful dramatic short "Take Flight," complete with green-screen effects and copious CGI. As Eric Kurland has noted, the Google Cardboard VR viewer is almost identical in purpose and design with Holmes's patent-free, mass-produced stereoscope from over 150 years ago,[165] although its 360-degree dimensionality makes it quite different in audiovisual terms.

165 Eric Kurland, "History of VR" in Celine Tricart, *Virtual Reality Filmmaking: Techniques and Best Practices in VR Filmmaking* (London: Routledge, 2018), 17. In 2019, The Sharper Image gift catalogue offered a sophisticated plastic version of the Google Cardboard Viewer with integrated Bluetooth earphones ($60), and a Pocket Video Drone to produce high-definition footage for it ($90).

At this point, enter the Oculus Rift, a product line of headsets manufactured by Oculus VR. Funded through a Kickstarter campaign, Oculus was a technology company founded by Palmer Luckey (b. 1992) in 2012 to produce consumer-friendly HMDs for gaming based on inexpensive smartphone parts.[166] When the company was acquired by Facebook (for $2 billion) in 2014, the headset was branded as the Oculus Rift, and CEO Mark Zuckerberg vowed to sell a billion of them by 2018. The Rift was first marketed to consumers in March 2016 at an initial cost of $600, while Oculus was simultaneously developing the Galaxy Gear VR headset with Samsung Electronics to work with its flagship smartphones. The original Rift required tethering to a PC to power it and used a low-resolution LCD display; subsequent versions, the Oculus Quest and Oculus Quest 2, were self-contained and had higher resolutions for their OLED displays. (In 2021, Oculus models retailed for between $300 and $400, depending on their processing power.)

In this context, VR was initially seen as a contribution to the gaming industry, whose worldwide revenues increased by 50 percent from 2012 to 2016 due to the rise of mobile gaming and the popularity of gaming in the Asia-Pacific market.[167] According to *The New York Times* in 2019, more than two billion people played video games, including 150 million Americans, nearly half the country's population, 60 percent of whom gamed daily. Video games were then one of the most lucrative sectors of the entertainment industry, overtaking film, television, music, and books. Games were simultaneously the most profitable kind of mobile application, accounting for about a third of all downloads.[168] Indeed, the proportion of income generated by mobile games grew dramatically from 10 percent in 2012 to 27 percent of the total market in 2016. The gaming industry was initially in the forefront of the VR consumer market development. For example, after the Samsung venture described above, Oculus announced a partnership with Microsoft to bundle the Rift with XBox One controllers, so that all XBox One games could be streamed to Oculus. This required Oculus to adopt Windows 10 as its operating system

---

166 The global success of the smartphone in the 2010s drove the quality of their high-resolution screens up and their costs exponentially down. Motion sensors and gyroscopes embedded in them could be appropriated by VR displays to track head, hand, and body positions, and the processing power of their chips was equal to that of former supercomputers. All of today's HMDs are built out of inexpensive smartphone technology; according to Kevin Kelly, "Put on almost any synthetic-reality display and you enter a world born of billions of phones" (Kevin Kelly, "Hypervision," *Wired* 24.05 [May 2016]: 78.)

167 Euengia Siapera, *Understanding New Media*, 2nd ed. (London: Sage, 2018), 233.

168 Ferris Jabr, "Can You Really Be Addicted to Video Games?," *New York Times*, October 27, 2019: 41.

and shows how closely VR was originally integrated with social media, games, software and hardware companies.[169]

As Graham Edwards has pointed out, however, many working in the expanding world of VR share a background in the mainstream film industry, often with experience in producing Visual Special Effects (VFX). They acknowledge the importance of bringing their skills not only to the making of VR but also to that of other media forms. Michael Berman, a former visual effects artists at Industrial Light & Magic and founder of Kaleidoscope VR, has said: "There are elements of films, games and theater. It's this kind of soup that everybody's splashing around in, making a big mess. It's a great time to do that, because the rules aren't set yet and people can just get in and have fun."[170] Traditional filmmakers can feel somewhat disoriented in the virtual realm. For example, if the camera sees everything where do you put the crew? Can you use conventional continuity editing? How can you direct an audience's attention when it is free to look anywhere it chooses? According to Patrick Meegan at Jaunt VR,

> You truly do need to think of this in a different way. Oftentimes, the first thing new filmmakers do when they come into 360-degree space is they put stuff *everywhere*. Instead, I look at it as being a choreographer—I want viewers to make the choice and look around, but I also want to guide them into a movement that feels good. It's like leading people in a subliminal dance.[171]

A good instance of this is the first venture into VR of Matthew Gratzner, creative director of New Deal Studios, who in mid-2014 directed *The Mission*—a seven-minute, 360-degree 3D video that placed viewers in the middle of a Second World War combat zone. Gratzner ignored the conventional 3D/VR wisdom to use only long takes with a locked-off camera and included a number of quick cuts and mobile camera shots with a Steadicam. "The so-called rules," he said, "really limit the ability to tell a story. For one shot, I had the camera attached to a real parachute that was 100 feet off the ground—we had a battle with 50 people [and] a full-scale tank."[172] A landmark in early VR filmmaking, *The Mission* employed actors with theater experience capable of maintaining performances and hitting multiple marks throughout relatively

169 Siapera, *Understanding New Media*, 234.

170 Quoted in Graham Edwards, "The Dreamsmiths Unleashed," *Cinefex*, 151 (February 2017): 82.

171 Ibid.

172 Ibid., 84.

long takes. Another early effort was Ray Tintori and Gabe Spitzer's 360-degree documentary *Follow My Lead*, centered on the 2016 NBA Finals. They deployed multiple cameras in the arena with an operator who captures moving shots using a gyroscopically stabilized monopod mount. Editors assembled the final 25-minute cut using Adobe Premiere with plug-ins that permitted them to work in a wraparound spherical environment, selectively rotating shots around a vertical axis—in effect, spinning around the entire space in order to align the focal point of each new shot with the focal point of the shot preceding it. Inspired by their sports documentary experience, the filmmakers used an impressionistic subjective editing style, thus minimizing the need for viewers to turn their heads to follow action.

Many new VR studios employ a classical Hollywood studio system model, producing and distributing their VR experiences, and also supplying their own camera equipment. The Jaunt ONE camera system, for example, illustrates the complexity of 360-degree live-action photography. The rig is comprised of 16 monoscopic cameras spaced around an equatorial mount, supplemented by four cameras each on the top and bottom. The resulting 24 video streams, recorded at 60 fps, are uploaded to a cloud-based renderer where stitching algorithms combine the individual files into a single HD master, using adjacent cameras to generate a three-dimensional image. In 2016, the Jaunt ONE was used on *Invisible*, a five-part VR serial about a family whose members carry a mysterious invisibility gene. Produced by Conde Nast, Jaunt, and Samsung and directed by Doug Limon (*Edge of Tomorrow*, 2014—a brilliant 2D to 3D conversion), the series contained many VSX which were handled in a two-stage process. First, they were applied to low-resolution "lat-longs," rectangular video files representing a flattened spherical view, in the same way that a two-dimensional world map represents spaces on a globe. Then, the lat-longs were mapped onto a sphere with a high-definition camera placed at the center for standard operations such as rotoscoping, painting, and applying filters. Finally, they were stitched back together into the original spherical space.

Such VFX challenges were common to all 360-degree films at the time, and software companies began to develop new toolsets to meet them, because the quality of the original stitch is fundamental to the whole process. According to Simon Robinson, chief scientist at The Foundry, "Stitching is hard for multiple reasons, especially on moving footage. You don't have a nodal camera, because there's an offset between all the physical cameras. So the convergence on foreground objects is different than the convergence on background objects."[173] Furthermore, VFX artists needed viable ways to manipulate 3D

173 Ibid., 86.

imagery on their 2D monitors until the tools were created that allow complex postproduction operations to be performed within a virtual environment. Moving creativity fully into the third dimension required filmmakers to enter the world of gaming: 360-degree films are prerendered, but more immersive VR experiences rely on game engines to render everything on the fly. Rather than simply gazing at the inner surface of a spherical screen on which scripted events run their course, the user enters a responsive 3D world where parallax shifts at the tilt of the head and he/she is free to explore the environment at will.

In 2016, Phil Tippet (b. 1951), master animator, visual effects supervisor, and founder of Tippet Studios, produced the immersive *Mad God VR* to accompany his surrealistic short film series *Mad God*. For it, stop-motion artists brought life to some 28 puppet characters that compositors inserted into a virtual 360-degree arena. Animators at Kaleidoscope reconfigured the images within the Unity game engine, placing characters on cards in depth so that viewers would experience credible parallax as they moved their heads around in space. Ultimately, VR specialists at Wevr made the production available to the public through its Transport delivery system.

At about the same time, Robert Stromberg (b. 1965), production designer for *Avatar* and director of *Maleficent*, founded the Virtual Reality Company (VRC), whose first major production was The *Martian VR Experience*, as an interactive ancillary to Ridley Scott's popular 3D feature *The Martian* (2015). In collaboration with several VFX houses, VRC positioned viewers in the shoes of Scott's stranded astronaut, permitting them to see the environment of Mars through the character's eyes and perform a number of narrative-driven tasks. VR experiences are often tied to the release of a major motion picture, for example, the 90-second *Battle for the Avengers Tower*, created by Framestore to promote *Avengers: The Age of Ultron* in 2015. Here, viewers are put in the place of Iron Man as he, Thor, Captain America, and Hulk grapple with Ultron Prime and his foot soldiers while missiles and other projectiles crisscross the spherical environment. In 2016, Lucasfilm and ILMxLAB released *Rogue One: Recon* as a complement to the film *Rogue One: A Star Wars Story*, while Framestore's *Fantastic Beasts VR* was released as a tie-in experience for *Fantastic Beasts and Where to Find Them*. Other VR licenses have involved Dreamworks product such as the *Shrek* and *Kung Fu Panda* franchises. (Perhaps the first such tie-in was *Ascend the Wall* which appeared as part of a 2014 *Game of Thrones* traveling exhibition—it allowed viewers to enter a model of a Black Wall elevator and be transported into the air over the fantasy land of Westeros.)

A major issue facing all VR creators, including those described above, is the question of presence: that is, should the inhabitants of a virtual world acknowledge the existence of the viewer? In some VR productions, a character will recognize the presence of the audience by making eye contact. This is the case with Oculus's *Henry*, the tale of a hedgehog who wants someone to hug him on his birthday, but at what point does this kind of interaction turn the movie into a video game?[174] This problem was confronted by Jon Favreau (b. 1966), producer and director of such digital 3D classics as the Iron Man series and *The Jungle Book*, which created a brilliant synthesis of live and animated 3D, when he created *Gnomes & Goblins*, whose story was scripted but whose design was more akin to that of a game. This tension in VR production between its status as cinema and game simultaneously was partially resolved when artists began to create VR inside of the virtual realm itself, a step initially taken by Oculus Story Studio when it produced *Dear Angelica*. A free-flowing, dreamlike story about a teenaged girl's memories of her actress mother, the film was fashioned from illustrations created entirely within the digital realm. The company's newly developed software tool, Quill, allowed graphic artist Wesley Allsbrook to paint his images directly in VR; rendered in real time, the resulting 3D artwork responds systematically to the viewer's gaze in the HMD.

Another company deeply involved in interactive storytelling is Oculus Story Studio whose mission is to harness the immersive potential of the Oculus Rift headset and which presented *Lost*, its first narrative experience, at the Sundance Film Festival in 2015. This envisions the plight of a detached mechanical hand as it attempts to find its way out of a moonlit forest. Other systems, however, allow full immersion not merely in fantasy realms but in photorealistic simulacra of the real world. Two companies, Lytro and Uncorporeal Systems, have built systems capable of live-action volume capture. For example, Uncorporeal's "videogrammetry" process surrounds a performer with as many as 48 synchronized cameras to derive data that can reconstruct the performer as a detailed digital double, which can then be inserted into a separate background and rendered by a game engine as a complete 360-degree environment. Pushing performance capture through a game engine allows artists to adjust depth of field, rack focus, surround sound cues, and lighting to keep the viewer's attention on the point of interest (POI).

Digital Domain has combined with branding agencies, network studios, and companies like Samsung, Facebook, and Google's YouTube to produce VR experiences, and it also delivers 360-degree live broadcasts, such as Oslo's 2016 Nobel Peace Prize Concert. For the Nike VR advertisement *The Neymar Jr. Effect*, Digital Domain allowed viewers to experience goal-scoring action from the viewpoint of an international soccer star. Compositors fashioned a digital stadium and filled it with live-action footage of the players, then deployed numerous camera rigs to capture to action from multiple points of

---

174 Ibid., 95.

view (POVs); in postproduction the VFX team blended that footage with CGI versions of the action across 360 degrees of the final image.

According to *Wired* magazine's founding executive editor, Kevin Kelly, VR does two important things. First, it generates an "intense and convincing" sense of presence, and second, it functions as real experience—not as the memory of something the viewer has seen but of something that has actually *happened* to him/her.[175] VR is, in the words of Jeremy Bailenson, a kind of "experience on demand" that "engulfs us."[176] As an example, Kelly describes a VR experience that he encountered in 2016 called the Void; it takes place in a large room and the user wears a 12-lb vest that carries batteries, a processor board, and 22 haptic patches that vibrate at the right moments. The headset isn't tethered so that the user is free to roam around the room to navigate an *Indiana Jones*-like adventure that appears to take place over a large territory, simulated by techniques known as "redirected walking" and "redirected touching" in which small sensory miscues are given to suggest a vast space in a 60- by 60-foot room that is matched by dynamic binaural sound.[177]

Ancillary to VR, which places the viewer in another location entirely, there are augmented reality and mixed reality. In the former, the visible natural world is overlaid by digital content; in the latter, virtual objects are integrated into the natural world and become responsive to it. Film director Peter Jackson remarks that none of these technologies are simply extensions of 3D movies but something completely different. "Once you can create the illusion of solid objects anywhere you want," he says, "you create new entertainment opportunities."[178]

As might be expected given the history of representational technologies, one such opportunity was sexual stimulation—not through virtual pornography but through *simulation*. In prior iterations, the technologies of photography, stereography, and motion pictures had been appropriated for pornographic purposes because they offered the increasingly realistic representation of sexual subjects. VR, however, offered the solidity of sexual experience itself—in other words, it achieved the Bazinian ideal not of representation

175 Kevin Kelly, "Hypervision," *Wired* 24.05 (May 2016): 78. Much the same was said of movies when they first appeared in the late nineteenth century, including the fear that audiences might mistake them for reality.

176 Jeremy Bailenson, *Experience on Demand: What Virtual Reality Is, How It Works, and What It Can Do* (New York: W. W. Norton, 2018).

177 Kelly write of these 3D acoustics: "To be persuasive, the apparent location of sound has to shift as you move your head. Deep presence includes the sensations of motion from your inner ear; if the two are out of synch with what you see, you get motion sickness" ("Hypervision," 80).

178 Quoted in "Hypervision," 85.

but "*re*-presentation"—not of showing things but creating the actual state of feeling them. (Compare the "feelies" of Aldous Huxley's *Brave New World* [1932].). In 2019 there were numerous companies that offered VR sexual simulations—for example, BaDoink, Wankz, and VR Bangers—including live stream "camming"—that is, performances by freelance sex actors and actresses live-shooting acts from sets and/or bedrooms in exchange for small micro-payments to their bank accounts or the sale of small cyber-tokens for one-on-one "Private Shows" (e.g., Chaturbate, Live Jasmin, and Cam4). As Peter Rubin noted in 2018: "There's gay VR porn, trans VR porn, BDSM VR porn; basically, find a flavor, stick 'VR' in the middle of it, and it likely exists—And if it doesn't, it soon will."[179]

For all of this, there have been no major VR installations since *Carne y Arena* in 2017, and VR has persistently failed to become a medium of mass entertainment. For example, VR is slated to become a main component of James Cameron's Avatar sequels but to date it is unclear how (although the original film has generated one successful Disney theme park ride). At this writing, VR still seems to be more the province of individual gaming behavior than anything approaching the tectonic industrial shift (however short-lived) to digital 3D in the wake of Cameron's 2009 windfall blockbuster, although a VR experience was included in the package for many of the "Immersive Van Gogh" exhibits—new, high-tech forms of the nineteenth-century panorama—that toured the United States and Canada, 2020–21.

179 Peter Rubin, "XXX-Change Program," in *Future Present: How Virtual Reality Is Changing Human Connection, Intimacy, and the Limits of Ordinary Life* (New York: HarperOne, 2018), 195–220.

Chapter 9

CONCLUSION

A great deal of human energy has been spent in trying to make flat surfaces look three dimensional. This is in part an effort to extend human vision with the promise that the depth perspective (and perhaps life itself) never ends.

From the foregoing, it should be clear that in one sense the history of three-dimensional cinema parallels, and sometimes burlesques, that of two-dimensional cinema—that is, it represents a kind of alternative current to the mainstream film industry (and, later, the gaming industry). When that industry began to experiment with sound and color during the 1920s, it also experimented with two-strip anaglyphic 3D; when it sought an audiovisual alternative to television in the 1950s, it briefly adopted two-strip polarized 3D; when it sought to exploit post-Code "porno-chic" and sensational violence in the wake of CARA in the 1970s and the 1980s, it turned to 3D again, this time in single-strip form; and finally when CGI and digital production/distribution transformed the industry globally in the 2000s, digital 3D became the blockbuster gold standard for a decade. Beyond that, VR clearly signals a new phase for 3D cinema, but it has failed to materialize as mass entertainment, although it has been mainstreamed in game-like formats for individual viewers.

The ideal of stereoscopy, however, has ancient roots. To see things "solid" on a flat surface has been the dream of science, art, and philosophy at least since the bas-reliefs of the Renaissance, which themselves had a long history in the art of ancient Egypt, Assyria, Babylon, and Rome. Systems for achieving this end realistically and their commercial exploitation go back to the seventeenth century, and the invention of photography in the mid-nineteenth century ensured that they would continue after the invention of motion pictures into the twentieth and twenty-first centuries.

The history of three-dimensional cinema then falls into several distinct stages. First there was the period experimentation in the formative years of the classical Hollywood system, 1915 (Porter and Waddle's Famous Players screenings) to 1951 (the Festival of Britain's Telecinema screenings). Then there was the Hollywood boomlet of 1952–55 during which the entire

American industry flirted with wholesale conversion to two-camera/two-projector "Three Dimension." Though it produced fewer than 70 features and was replaced by single-camera/single-projector anamorphic widescreen (versions of Fox's CinemaScope) and, later, non-anamorphic 70mm systems, this phase of 3D produced some optically superior motion pictures that were ill-served by occasional parallax distortion and shoddy projection practices. Stereoscopic cinema was kept alive for the next few decades by single-lens over-and-under (and, sometimes, side-by-side) systems like Optimax that trafficked in exploitation genres such as pornography and horror, and then by large-format IMAX documentaries. During the 2000s, the arrival of digital cinema offered cost-effective and optically satisfying digital 3D using single cameras and projectors. The exponential box-office and critical success of James Cameron's *Avatar* in 2009 opened the floodgates for digital 3D, yielding more than 600 three-dimensional features worldwide, nearly half of them computer-animated, over the next decade; by 2020 the process lost its audience allure and was largely replaced by 2K and 4K digital projection both theatrically and at home—at least in Western markets. (Digital 3D, however, remains wildly popular in Asia-Pacific territories and in parts of the Middle East and Latin America.)

For nearly a decade after *Avatar*, digital 3D became a form of cultural LSD slipped into the theatrical entertainment water supply. Audiences would pay a significant upcharge to see even the worst of it, and literally hundreds of billions of dollars exchanged hands in the production, release, and consumption of meretricious blockbusters based on comic books, video games, and retrofitted fairy tales. In their homes, people watched hundreds of hours of post-converted 2D movies and sporting events on 3D TVs that were never meant to be seen that way.[180] Some big-budget films were made during the period that used 3D in innovative and even necessary ways (e.g., Martin Scorsese's *Hugo* [2011], Alfonso Cuaron's *Gravity*, [2013]), but most still had "gimmick" written all over them.

The future of 3D in three films

Billy Lynn's Long Halftime Walk (2016) is a serious film about the Iraq war and its domestic consequences, adapted from a novel, and shaped as commercial entertainment using state-of-the-art 3D and HFR technology. *They Shall Not Grow Old* (2018) is an experimental archival documentary that uses 3D

180 Confession: I myself watched Blu-ray editions of the *Alien* Quatrilogy, *The Innocents*, and *Citizen Kane* post-converted on my 3D TV with great pleasure.

conversion as part of its restoration arsenal. *Long Day's Journey into Night* (2018) is a two-dimensional art film that employs 3D in its final hour to represent a dream state or wish fulfillment fantasy. Their genres could not be more different from an historical/marketing standpoint, and they each point in the direction of a future for three-dimensional cinema.

Billy Lee's Long Halftime Walk was adapted from a 2012 novel of the same name by Ben Fountain. It was directed by Ang Lee at a production cost of $40 million because DP John Toll shot it in 3D at the extra-high frame rate of 120 fps in 4K resolution (a first for any feature). Since only five theaters worldwide were equipped to show the film at its maximum frame rate and highest resolution, additional versions were prepared for its wide release, including 120 fps in 2D and 60 fps and 24 fps in 3D. For its presentation at the 54th New York Film Festival, two Christie Mirage 4K laser projectors were installed together with the RealD 3D system's 42-by-19-foot Ultimate Screen technology that displayed 28-ft-L per eye. *Billy Lynn*'s *Long Halftime Walk* was a commercial failure due to both its complex release format and its subject matter. Billy Lynn is a 19-year-old US Army specialist from Texas who is caught on a cable news-crew camera dragging his sergeant to safety during an intense firefight in Iraq (the sergeant later dies). Lynn is awarded a Silver Star, and he and the members of his "Bravo Squad" unit become national media celebrities and are sent on a publicity tour that ends in a gala halftime show with "Destiny's Child" at the Dallas Cowboys November 25 Thanksgiving home game. During the course of this gig, Lynn experiences multiple flashbacks to the war zone and his homecoming that reveal deep cynicism about his team's mission and the American presence in Iraq (Figure 9.1). The media tour itself shows the ordinary citizens' blithe ignorance of the nature of the war and gross insensitivity toward the sacrifices that troops like those of "Bravo Squad" are making on their behalf. Serious reflections on the Iraq war were not good box office in 2016, nor was the complicated structure of a film that moved back and forth among several time frames and included occasional flashes of absurdist humor.

Lee purportedly wanted the extra high frame rate and high-resolution 3D to make the combat sequences immersive and hyperrealistic to the audience, but for some critics they had the opposite effect. David Rooney of *The Hollywood Reporter*, who praised the film as "a highly original character study," also said that "the technical innovations took me out of the drama just as often as they pulled me in." And like *The Hobbit: An Unexpected Journey* (Peter Jackson, 2012), which used a high frame rate of 48 fps, *Billy Lynn* was criticized for having an unpleasant shot-for-video look. For example, Dan Callahan writing for *TheWrap* said that some of the characters were "so superclear that they look like a cut-out with scissors from a glossy magazine" and said "The extra-clarity 3D in the Lee movie often looks weirdly like something shot

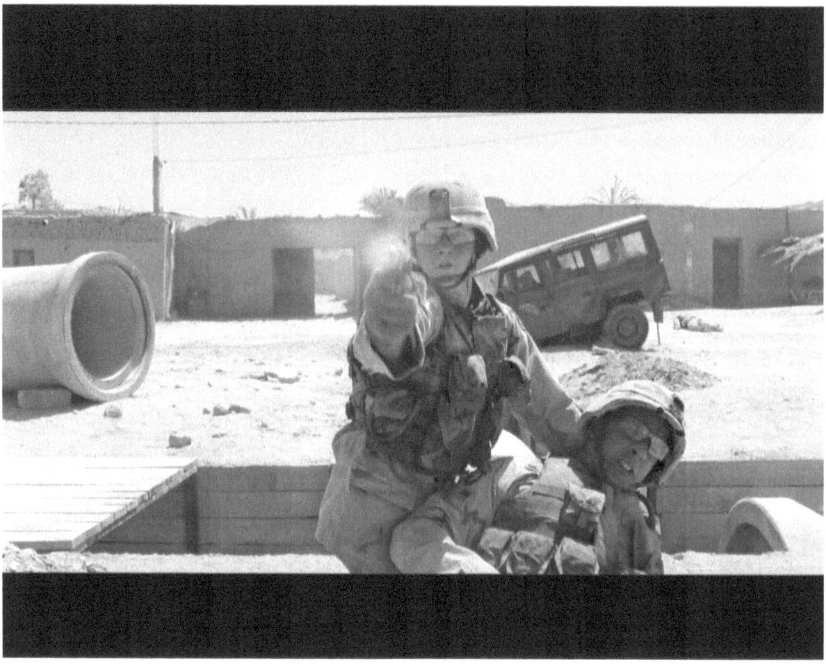

Figure 9.1 *Billy Lynn's Long Halftime Walk* (Sony Pictures Releasing, 2016): shot with the Sony CineAlta F65 System for presentation at several different frame rates and resolutions in RealD Cinema. Billy, in combat, fires his gun directly into the camera—an old trick used in cinema well before 3D (and as early as Edwin S. Porter's 1903 *The Great Train Robbery*); see VR quote, p. 138.

on videotape in the 1980s."[181] It seemed clear to both audiences and critics alike that the movie didn't need 3D or even HFR to make an impact, and it returned only $30.9 million globally. (It earned $20 million of that in China, where Lee and 3D are both very popular, even though domestic audiences were unable to relate to the Iraq war experience of its central character.) So, *Billy Lynn's Long Halftime Walk* is a film that failed to justify its high-end budget on the three-dimensional technical side, and in theaters it probably would have played just as well (or poorly) flat at 24 fps. The preferred version on Blu-ray is the 4K Ultra HD running at 60 fps, which is optically superior to the 24-fps 3D version.

181 https://en.wikipedia.org/Billy_Lynns_Long_Halftime_Walk/film. Accessed January 7, 2020.

In *They Shall Not Grow Old* Peter Jackson created a truly unique feature-length documentary (theatrical version, 99 minutes; extended cut 129 minutes) immersing the audience in the experience of the average British soldier on the Western Front of the First World War from 1914 to 1918, using archival footage taken at the time and expertly colorized and edited; the footage derives from the Imperial War Museum, which coproduced the film with the BBC. There is no narration and no historical contextualization. Instead, we hear the voices of veterans interviewed by the BBC in the 1960s and 1970s lip-synched into the restored footage, which does not begin until 25 minutes into the film. The first images are of modified 2.5 D black-and-white footage positioned at mid-screen within a small academy frame, which then expands into its full-color, widescreen, stereoscopic format for the duration until a few minutes before the film's conclusion. The film recounts both the horrors of trench warfare and the absurdities of daily life in the field, all told with the rough humor of the men who lived them. The film is dedicated to Jackson's grandfather, who fought in the war, and he received no fee for directing it. (In fact, Jackson's WingNut Films visually restored *all* of the footage the Imperial War Museums sent them at no cost in order to upgrade the archive.)

Jackson sorted through 100 hours of original footage and 600 hours of audiotape from interviews with 200 veterans, 123 of whose voices appear in the film. The colorization was achieved through meticulous research, with great accuracy given to the hues of the British and German uniforms and battlefield landscapes. Most of the 3D effects are immersive and occur in positive parallax, although there are a number of visual elements that protrude toward the screen. Exploding shells, firing cannons, and horse-driven vehicles all come toward the camera but never break the window of the screen (Figure 9.2).

In fact, the film is noted for its restrained use of 3D to separate persons and objects from their environments, which themselves have considerable depth and increase the realism of the presentation. Reviews were almost universally positive, with *The Globe and Mail* calling it "The best and most warranted use of 3D today";[182] *The Guardian* said it was "a visually staggering thought experiment";[183] *Home Theater Forum* said that the film was "The 3D event of the year," lending astonishing cinematic intimacy to life during the First World War: "Jackson does more than restore archival footage; he has restored the humanity of men caught up in one of history's great cataclysms,"[184] while

182 www.theglobeandmail.com/film/1918. Accessed January 7, 2020.
183 www.theguardian.com/film/1918. Accessed January 14, 2020.
184 www.hometheaterforum.com/they-shall-not-grow-old. Accessed February 2, 2020.

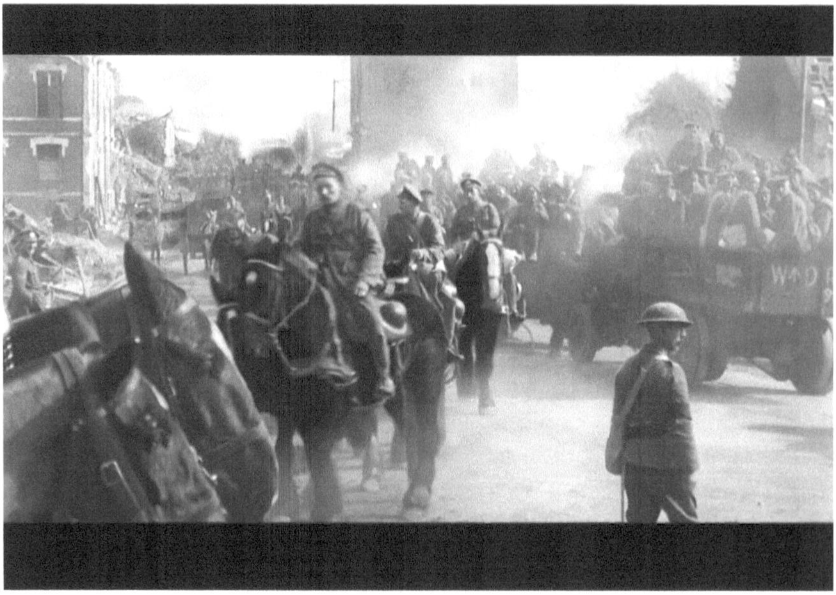

Figure 9.2 *They Shall Not Grow Old* (Warner Bros. Pictures, 2018): shot in Spherical 35mm matted to 1.85:1, colorized, and converted to 3D and Dolby Digital Sound in postproduction. This wide shot displays nearly perfect convergence—the horses come toward the camera from the rear but never break the window of the screen.

Popular Threads said "One of the most amazing aspects of it is the excellent 3D conversion done on most of the film to make it even more immediate to viewers" and *CinemaBlend* called it "a classic case of using modern technology to give the audience a fresh perspective on history."[185] Yet it would be wrong to say that 3D is essential to *They Shall Not Grow Old*; in fact, Jackson doesn't even mention the meticulous 3D conversion by Stereo D in the "Making of" documentary on the Warner Bros. Blu-ray disc, which is devoted mainly to the digital cleanup and colorization of the archival footage and the use of accurate acoustic and dialogue effects. Furthermore, the Blu-ray disc of the film was not released in 3D in any markets (a response, no doubt, to the eclipse of 3D TV), so the only way to see it in stereo was in a theater. Yet, *They Shall Not Grow Old* is definitely enhanced by 3D and the three-dimensional version is by far the preferred one. It demonstrates how 3D can be used in an original way to engage an audience with material long thought to be irredeemably

185 www.cinemablend.com/reviews. Accessed February 3, 2020.

archaic, and it proved popular enough to earn $20.4 million worldwide, a considerable sum for a documentary feature.

Long Day's Journey into Night (no relation to the Eugene O'Neill play of the same name: the original Mandarin title is *Last Evenings on Earth*) was directed by Bi Gan in 2018 as a self-conscious art film—in fact, it has been called "the most successful art film hit in Chinese cinema."[186] It tells the noirish story of Lu Hongwu, who has returned to his hometown of Kaili in Guizhou province, where the film was shot on location, for his father's funeral. There he finds an old photograph of his long-lost love Wan Qiwen and begins a melancholy search for her that features flashbacks to the summer of 2000 when they had an affair and she was the girlfriend of a ruthless local gangster. As he wanders around the dilapidated and run-down city of Kaili—a landscape of desolation, despair, and hopelessness—he encounters various eccentric characters and eventually ends up in a seedy movie house where he puts on 3D glasses (as the film's real theatrical audience is intended to do) to watch a film call "Long Day's Journey into Night."

This second half of the film unfolds in one continuous 59-minute 3D take, shot with an Arri Alexa RED camera, that Bi wanted to feel like a "lucid dream." During the long take, which Peter Travers of *Rolling Stone* called "the mind-blowing 3D shot of the century,"[187] the camera floats from the interior of a cave, to a motorcycle, to a zipline and finally takes off on a drone, a sequence shot on the RED camera in 2D and then converted to 3D in postproduction. In the course of his three-dimensional dream, Lu meets a woman who strikingly resembles Wan Qiwen (and is in fact played by Tang Wei, the same actress) (Figure 9.3). After a series of remarkably well-choreographed misadventures, Lu finally realizes that he has found his lost love, and they kiss passionately while sitting on the bed of an abandoned house, with the room spinning around them in an image similar to the "recognition" scene in Hitchcock's *Vertigo*.[188] While it took seven full attempts to get the 59-minute shot right, preparation was the biggest battle—months of experimentation leading up to

186 David Sims. "The Mind-Boggling Film That Broke China's Box Office," www.theatlantic.com/entertainmnet/archive. The reason for this was that the film was marketed as a romantic melodrama and a perfect date movie for its New Year's Eve release; by December 25, *Long Day's Journey into Night* had presold $15 million in tickets, and it grossed another $26 million in its first three weeks (most of that from its premiere). When word-of-mouth got out on social media that it was an art film, its box office plummeted; a dismal French release yielded another $ 1 million.

187 Quoted from the cover of the Kino Lorber Blu-ray edition.

188 In this respect *A Long Day's Journey* resembles both Lou Ye's *Suzhou River* (2000) and Christian Petzold's *Phoenix* (2014).

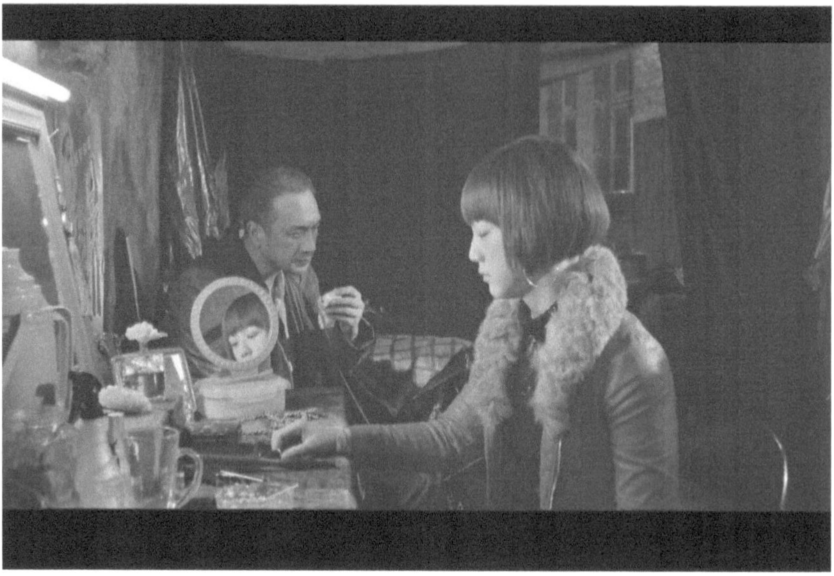

Figure 9.3 *Long Day's Journey Into Night* (BAC Films, 2018): 3D long take shot with Arri Alexa Red Helium Camera system and Zeiss Super Speed Lenses for presentation in RealD Cinema at 1.85:1. A frame enlargement from a scene near the end of the 58-minute long take in the second part of the film. Lu Hongwu (Jue Huang) and the putative Wan Qiwen (Tang Wei) in her temporary dressing room.

the shoot meant that Bi had to shut down production twice in an attempt to get it right.

In interviews, Bi has said,

> Godard [*Goodbye to Language*, 2014] and Wenders [*Pina*, 2011] and Cuaron [*Gravity*, 2013] did 3D films that really inspired me that the form could be used for art films. But I feel that in history 3D is almost like [...] half a movie? So most of it, when they use 3D it's that effect of pushing or popping out at you so you're feeling something kinetic. And I did not want to go there is this film. The only time you feel something like that in my 3D environment is with faces. That's the only time I wanted to allow the audience to feel the depth. Other than that, everything is left to the background as just part of the environment. [...] I'm not really interested in 3D as a way of shooting films. [...] But what I am interested in is how it feels almost like [...] Memory. When you close your eyes and remember, you almost seem to see it in a very three-dimensional way, even though it's something recalled from the past, something that's already happened. I feel that when you open your eyes and you're back

> into this world and the present, that 3D was the best way to capture that sensation of the difference between memory and existing in the present. We spent a lot of time researching the history of 3D, examining what kind of effect filmmakers were using over time to make these types of films. One of the biggest things was how to connect different shots in 3D, how to not make it jarring to go a from a long lens close-up to a wide shot. [...] And maybe that's why a lot of 3D movies aren't good. [...] The way our eyes see things in 2D and 3D at the same time is so complicated, that even exporting the DCP [Digital Cinema Package for distribution] was so difficult, technically, to get the final shot working."[189]

To heighten the sense of 3D space and color, Bi incorporated a Dolby Atmos mix into the sound design of the second half: "The film becomes even more immersive, just because of the dimensional changes, visually, with the 3D. You will have that one extra layer of dimensional audio sound as well. So that's really going from a 2D flat, to this 3D immersive and all the surrounding audio environment I wanted."[190]

To sum it all up, Glenn Kenny writing for *The New York Times* said,

> Not since David Lynch's *Inland Empire* has a filmmaker in such proximity to what we can call the independent mainstream undertaken quite as radical a challenge to linear narrative [and] the achievement is all the more awe-inspiring given that the movie's second half, a 3-D film-within-a-film (of sorts; it could just be Luo's dream of one) is contained within what seems to be a single continuous shot of nearly an hour in length, [...] [building up] to a moment of breathtaking romanticism that's as intoxicating as it is unexpected.[191]

From these analyses, it seems clear that the future of three dimensional cinema lies not in the standard dramatic feature, no matter how immersive its action scenes, but in artful experimentation, where its full integration into films like *They Shall Not Grow Old* and *Long Day's Journey into Night* can be a positive aesthetic value rather than a gimmicky add-on to enhance a film's market

189 Aaron Stewart-Ahn, "A Half-Year Interview with Bi Gan, in Two Parts," Booklet accompanying the Kino Lorber Blu-ray edition of *Long Day's Journey into Night* (2019).

190 Chris O'Falt, "How the Gravity-Defying 59-Minute 3D Take in 'Long Day's Journey into Night Was Shot'," www.indiewire.com.

191 Glenn Kenny, "A Dark and Mesmerizing Associations," *New York Times*, April 12, 2019: C8.

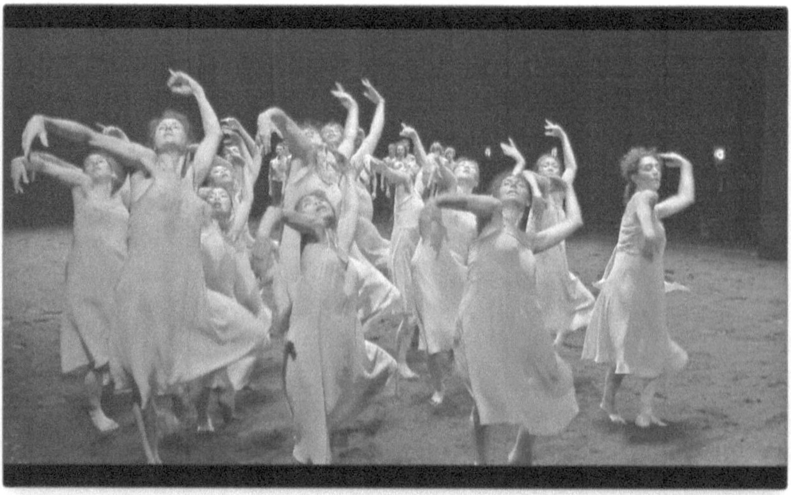

Figure 9.4 *Pina* (HanWay Films, 2011): shot with the Sony CineAlta HDC-1500 camera system and Zeiss DigiPrime Lenses for presentation in RealD Cinema at 1.85:1. Tina Bausch's modernist dance troupe performs in the Tanztheater Wuppertal style in one of several films that points the way to the future of three-dimensional cinema.

appeal.[192] Such films would include Werner Herzog's *Cave of Forgotten Dreams* (2010), a documentary rendition of the oldest human-painted images on the walls of the Chauvet Cave in southern France, which uses 3D to capture the contours of the stone as they shaped the paintings; *Pina* (2011), Wim Wenders's documentary tribute to the contemporary dance choreographer Pina Bausch; and Jean-Luc Godard's *Goodbye to Language* (2014), a radical interrogation of the stereoscopic process in cinema and of binary vision itself (Figure 9.4).

Now that the mass audience has done with "drinking the [3D] Kool Aid" maybe it's time for avant-garde and experimental uses of 3D to hold sway once again, as they did at the form's beginning, when Eisenstein saw it as inevitable in experimental terms as the sun rising in the morning. The ideal of a fully three-dimensional cinema will never disappear because André Bazin has told us that it will not. The iconic French film theorist believed that each new technological development in the history of cinema (e.g., the introduction

192 The idea of adding stereoscopy to restored documentary content goes back at least to the Lumiere's 1935 3D conversion of *Arrivee d'un train en gare* 1895), and using 3D to represent dream states or hallucinations begins with *The Mask* (Julian Roffman, 1961).

of sound, color, widescreen, and 3D) brought it closer to its ultimate destiny—the production of a total simulacrum of reality—which was implicit from its very beginnings. As he wrote in "The Myth of Total Cinema" in 1946:

> The guiding myth, the inspiring the invention of cinema, is the accomplishment of that which dominated [...] all the techniques of the mechanical reproduction of reality in the nineteenth century, from photography to the phonograph, namely an integral realism, a recreation of the world in its own image, an image unburdened by the freedom of interpretation of the artist or the irreversibility of time. [...] Every new development added to the cinema must, paradoxically, take it nearer and nearer to its origins. In short, cinema has not yet been invented![193]

Had Bazin lived to see *Pina*, *They Shall Not Grow* Old, or *Long Day's Journey into Night*, he might disagree with that last sentence, but had he sampled most of the nearly 600 3D films produced internationally between 2010 and 2020, I don't think he would much change his opinion.

---

193 "The Myth of Total Cinema," in André Bazin, *What Is Cinema?*, vol. 2, trans. Hugh Gray (Berkeley, CA: University of California Press, 1967).

3D DISCOGRAPHY: DISCS VIEWED OR SAMPLED IN PREPARATION FOR THIS BOOK (AND MOST STILL AVAILABLE AT REASONABLE PRICES FROM ONLINE SOURCES)

Chapter 1

Stereographs: 3D Portals to the Past Presents Episode One: Photographs from the American Civil War Era, Volume One in 3-Dimensions (A Glenn Bonat Production from Helicon Digital Arts, 2017). 3D Blu-ray. 165 min. Original stereography 1861–65.

Around the Word in Three Dimensions. Stereographs: 3D Portals to the Past Episode Three (A Glenn Bonat Production from Helicon Digital Arts, 2020, directed by Glenn Bonat). 3D Blu-ray. 245 min. Original stereography from the late 1800s to the early 1900s.

Chapter 2

3-D Rarities, 1922–1962 (Flicker Alley/3-D Film Archive, 2015). 3D Blu-ray. 147 min. Original releases 1922–62.

Chapter 3

House of Wax (Warner Bros., 2013). 3D Blu-ray. 88 min. Original release 1953.

Man in the Dark (Twilight Time, 2014). 3D Blu-ray. 70 min. Original release 1953.

It Came from Outer Space (Universal, 2016). 3D Blu-ray. 80 min. Original release 1953.

Sangaree (Kino Lorber/3-D Film Archive, 2018). 3D Blu-ray. 95 min. Original release 1953.

Jivaro (Kino Lorber/3-D Film Archive, 2019). 3D Blu-ray. 92 min. Original release 1954.

The Maze (Kino Lorber/3-D Film Archive, 2018). 3D Blu-ray. 88 min. Original release 1953.

Dragonfly Squadron (Olive Films, 2014). 3D Blu-ray. 83 min. Original release 1954.

Wings of the Hawk (Kino Lorber/3-D Film Archive. 3D Blu-ray. 81 min. Original release 1953.

Inferno (Twilight Time, 2017). 3D Blu-ray. 84 min. Original release 1953.

Those Redheads from Seattle (Kino Lorber/3-D Film Archive, 2017). 3D Blu-ray. 90 min. Original release 1953.

Kiss Me Kate (Warner Bros., 2015). 3D Blu-ray. 110 min. Original release 1953.

Hondo (Paramount, 2012). 2D Blu-ray. 84 min. Original release 1953.

Gun Fury (Twilight Time, 2017). 3D Blu-ray. 82 min. Original release 1953.
Miss Sadie Thompson (Twilight Time/3-D Film Archive). 3D Blu-ray. 91 min. Original release 1953.
Cease Fire (Kino Lorber/3-D Film Archive, 2017). 3D Blu-ray. 78 min. Original release 1953.
Taza, Son of Cochise (Kino Lorber/3-D Film Archive, 2020). 3D Blu-ray. 79 min. Original release 1954.
Creature from the Black Lagoon (Universal Studios, 2013). 3D Blu-ray. 80 min. Original release 1954.
The Mad Magician (Twilight Time/3-D Film Archive, 2017). 3D Blu-ray. 72 min. Original release 1954.
Dial M for Murder (Warner Bros., 2012). 3D Blu-ray. 105 min. Original release 1954.
Gog (Kino Lorber/3-D Film Archive, 2016). 3D Blu-ray. 83 min. Original release 1954.
The Revenge of the Creature (*Die Rache des Ungeheurs* [German version]; Koch Media, 2015). 3D Blu-ray (anaglyphic). 82 min. Original release 1955.
3-D Rarities, Volume II, 1941–1983 (Flicker Alley/3-D Film Archive, 2020). 3D Blu-ray. 153 min. Original releases 1941–83.

Chapter 4

September Storm (Kino Lorber/3-D Film Archives, 2016). 3D Blu-ray. 92 min. Original release 1960.
The Mask (Kino Lorber/3-D Film Archive, 2015). 3D Blu-ray. 83 min. Original release 1961.
The 3-D Nudie Cuties Collection (Kino Lorber/3-D Film Archive, 2019). 3D Blu-ray. 166 min. Original release 1962; 1953; 1951.
The Bubble (Kino Lorber/3-D Film Archive, 2014). 3D Blu-ray. 91 min. Original release 1966.
The Stewardesses in 3D (Kino Lorber/Jezebel, 2016). 3D Blu-ray. 93 min. Original release 1969.
La marca del hombre lobo (*The Mark of the Werewolf*; DC Films, 1969). 2D DVD. Original release 1969.
Flesh for Frankenstein (Criterion, 1998). 2D DVD. 95 min. Original release 1973.
*Ape 3-D Aka A*P*E* (Kino Lorber/3-D Film Archive, 2017). 3D Blu-ray. 87 min. Original release 1976.
Parasite 3-D (Kino Lorber/3-D Film Archive, 2019). 3D Blu-ray. 85 min. Original release 1982.
Friday the 13th—Part 3: 3D (Paramount Home Entertainment). 3D Blu-ray (anaglyphic). 95 min. Original release 1983.
Jaws 3-D (Universal Pictures, 2016). 3D Blu-ray. 99 min. Original release 1983.
Amityville 3-D (Shout! Factory, 2013). 3D Blu-ray. 105 min. Original release 1983.
Silent Madness (Vinegar Syndrome, 2020). 3D Blu-ray. 92 min. Original release 1984.
Metalstorm: The Destruction of Jared-Syn (Shout! Factory, 2016). 84 min. Original release 1983.
Comin at Ya! (MVD Visual, 2016). 3D Blu-ray. 91 min. Original release 1981.
Treasure of the Four Crowns (Shout! Factory, 2014). 2D DVD. 97 min. Original release 1983.
Dynasty (Kino Lorber/3-D Film Archive, 2021). 3D Blu-ray. 92 min. Original release 1977.

Chapter 5

Captain EO (Antonio-MJJ Productions, 1989, 2005). 2D DVD. 17 min. Original release 1986.
Legends of Flight (Image Entertainment, 2011). 3D Blu-ray. 45 min. Original release 2010).

IMAX Hubble 3D (Warner Home Video, 2013). 3D Blu-ray. 44 min. Original release 2010.

The Universe: 7 Wonders of the Solar System (A&E Entertainment, 2011). 3D Blu-ray. 47 min. Original release 2010.

Rescue 3D (Image Entertainment, 2012). 3D Blu-ray. 46 min. Original release 2011.

The Last Reef: Cities Beneath the Sea (Shout! Factory, 2016). 44 min. Original release 2012.

Best of 3D: The Ultimate Collection (Universal Pictures, 2013). 3D Blu-ray. 85 min. Original release 2013.

To the Arctic (Warner Home Video, 2013). 3D Blu-ray. 40 min. Original release 2012.

Wonders of the Arctic (Shout! Factory, 2014). 3D Blu-ray. 44 min. Original release 2014.

Flight of the Butterflies (Shout! Factory, 2016). 3D Blu-ray. 44 min. Original release 2013.

The Polar Express (Warner Bros., 2004). 3D Blu-ray. 100 min. Original release 2004.

Tim Burton's The Nightmare before Christmas—20th Anniversary Edition (Walt Disney Studios Home Entertainment, 2003). 3D Blu-ray. Original 2D release 1993.

Beauty and the Beast (Walt Disney Studios Home Entertainment, 2011). 3D Blu-ray. 84 min. Original 2D release 1991.

The Lion King (Walt Disney Studios Home Entertainment, 2011). Blu-ray 3D. 88 min. Original 2D release 1994.

Top Gun: Limited 3D Edition (Paramount Pictures, 2013). Blu-ray 3D. 110 min. Original 2D release 1986.

Titanic: Limited 3D Edition (Paramount Pictures, 2012). Blu-ray 3D. 195 min. Original 2D release 1996.

Jurassic Park (Universal Studios Home Entertainment, 2013). Blu-ray 3D. 126 min. Original 2D release 1993.

Terminator 2: Judgement Day (Universal Studios Home Entertainment, 2017). 137 min. Original 2D release 1991.

The Wizard of Oz (Warner Bros., 2013). 3D Blu-ray. 101 min. Original 2D release 1939.

Spy Kids 3-D: Game Over (2003)/*The Adventures of Shark Boy and Lava Girl in 3-D* (2005) (Miramax Lionsgate, 2011). Blu-ray 3D. 84 and 89 min. Original 3D anaglyphic releases in 2003 and 2005, respectively.

Chicken Little (Walt Disney Animation, 2011). 3D Blu-ray. 81 min. Original release 2005.

Ghosts of the Abyss (Walt Disney Studios Home Entertainment, 2012). 3D Blu-ray. 92 min. Original release 2003.

Monster House in 3D (Columbia Pictures, 2010). 3D Blu-ray. 91 min. Original release 2010.

Beowulf 3D (Warner Bros./Film Per Tutti, 2014). 3D Blu-ray. 114 min. Original release 2007.

Meet the Robinsons (Walt Disney Animation, 2011). 3D Blu-ray. 94 min. Original release 2007.

Scar 3D (Phase4Films, 2011). 3D Blu-ray. 78 min. Original release 2007.

Journey to the Center of the Earth (New Line Cinema, 2008). 93 min. Original release 2008.

Bolt 3D (Walt Disney Studios Home Entertainment, 2011). 91 min. Original release 2008.

My Bloody Valentine 3D (Liongate, 2009). 3D Blu-ray. 103 min. Original release 2009.

Coraline (Universal Pictures Home Entertainment, 2011). 3D Blu-ray. 100 min. Original release 2009.

Monster vs. Aliens (20th Century Fox Home Entertainment, 2009). 3D Blu-ray. Original release 2009.

Up (Disney/Pixar, 2010). 3D Blu-ray. 96 min. Original release 2009.

Ice Age: Dawn of the Dinosaurs (20th Century Fox Animation, 2009). 3D Blu-ray. Original release 2009.

Final Destination 3-D (Warner Bros., 2010). 3D Blu-ray. 128 min. Original release 2009.

Dark Country (Sony Pictures Home Entertainment, 2009). 3D Blu-ray. 88 min. Original release 2009.

Cloudy with a Chance of Meatballs (Columbia Pictures, 2009). 3D Blu-ray. 90 min. Original release 2009.

Disney's A Christmas Carol (Walt Disney Studios Home Entertainment, 2010). 3D Blu-ray. Original release 2009.

Avatar (20th Century Fox, 2012). 3D Blu-ray. 162 min. Original release 2009.

Chapter 6

Post-2009 Digital 3D, mainly live-action

The Last Airbender (Paramount Pictures/Nickelodeon, 2010). 3D Blu-ray. 103 min. Original release 2010.

Alice in Wonderland (Walt Disney Studios Home Entertainment, 2010). 3D Blu-ray. 108 min. Original release 2010.

Step up 3D (Touchstone Home Entertainment, 2010). 3D Blu-ray. 107 min. Original release 2010.

Despicable Me (Universal Pictures, 2010). 3D Blu-ray. 95 min. Original release 2010.

Legend of the Guardians: The Owls of Ga'Hoole (Warner Bros., 2010). 3D Blu-ray. 97 min. Original release 2010.

Resident Evil: Afterlife (Sony Pictures/Screen Gems). 3D Blu-ray. 97 min. Original release 2010.

Clash of the Titans (Warner Bros., 2010). 3D Blu-ray. 108 min. Original release 2010.

Wrath of the Titans (Warner Bros., 2012) 3D Blu-ray. 99 min. Original release 2012.

The Green Lantern (Warner Bros., 2011). 3D Blu-ray. 114 min. Original release 2011.

Captain America: The First Avenger (Paramount Home Media Distribution, 2011). 3D Blu-ray. 124 min. Original release 2011.

The Green Hornet (Columbia Pictures, 2011). 3D Blu-ray. 119 min. Original release 2011.

Hugo (Paramount Home Entertainment, 2012). 3D Blu-ray. 126 min. Original release 2011.

The Adventures of Tintin: The Secret of the Unicorn (Paramount Home Entertainment, 2012). 107 min. Original release 2011.

BFG: Sophie & der Riese (Amblin Entertainment/Constantin Film, 2016). 3D Blu-ray (German version). 117 min. Original release 2016.

Twixt (20th Century Fox Home Entertainment, 2013). Blu-ray. 90 min. Original release 2011.

The Life of Pi (20th Century Fox Home Entertainment, 2013). 3D Blu-ray. 127 min. Original release 2012.

Billy Lynn's Long Halftime Walk (Sony Pictures Home Entertainment, 2017). 3D and 4K Ultra HD Blu-ray. 113 min. Original release 2016.

Prometheus (20th Century Fox Home Entertainment, 2012). 3D Blu-ray. 124 min. Original release 2012.

Exodus: Gods and Kings (20th Century Fox Home Entertainment, 2015). 3D Blu-ray. 150 min. Original release 2014.

The Martian (20th Century Fox Home Entertainment, 2016). 3D Blu-ray. 141 min. Original release 2015.

Blade Runner 2049 (Warner Bros., 2018; also distributed in Oculus VR format). 3D Blu-ray. 164 min. Original release 2017.

The Great Gatsby (Village Roadshow Pictures/Warner Bros., 2013). 3D Blu-ray. 142 min. Original release 2013.

The Hobbit: An Unexpected Journey (Extended Edition; MGM/Warner Bros., 2013). 3D Blu-ray. 182 min. Original release 2012.

The Hobbit: The Desolation of Smaug (Extended Edition; MGM/Warner Bros., 2014). 3D Blu-ray. 186 min. Original release 2013.

The Hobbit: The Battle of the Five Armies (Extended Edition; MGM/Warner Bros., 2015). 3D Blu-ray. 164 min. Original release 2014.

They Shall Not Grow Old (Warner Bros., 2018). Blu-ray. 99 min. Original release 2018.

How to Train Your Dragon (Dreamworks Home Entertainment, 2014). 3D Blu-ray. 98 min. Original release 2010.

Piranha 3D (Sony Pictures Home Entertainment, 2011). 3D Blu-ray. 88 min. Original release 2010.

Drive Angry (Summit Entertainment, 2011). 3D Blu-ray. 104 min. Original release 2011.

Priest (Sony Picture/Screen Gems, 2011). 3D Blu-ray. 87 min. Original release 2011.

Sanctum 3D (Universal Studios Home Entertainment, 2011). 3D Blu-ray. 109 min. Original release 2011.

Shrek Forever After (Paramount Pictures, 2010). 3D Blu-ray. 93 min. Original release 2010.

The Chronicles of Narnia: The Voyage of the Dawn Treader (20th Century Fox Home Entertainment, 2011). 3D Blu-ray. 115 min. Original release 2010.

Pirates of the Caribbean: On Stranger Tides (Jerry Bruckheimer Films/Walt Disney Pictures, 2011). 3D Blu-ray Limited Edition. 137 min. Original release 2011.

Iron Man 2 (Marvel Studios/Paramount Home Media Distribution, 2010). Blu-ray. 125 min. Original release 2010.

Thor (Marvel/Paramount Home Media Distribution, 2013). 3D Blu-ray. 114 min. Original release 2011.

The Avengers (Marvel Studios/Walt Disney Studios Home Entertainment, 2012). 3D Blu-ray. 143 min. Original release 2012.

Iron Man 3 (Marvel Studios/Walt Disney Studios Home Entertainment, 2013) Blu-ray 3D. 125 min. Original release 2013.

Thor: The Dark World (Marvel Studios/Walt Disney Studios Home Entertainment, 2014). 3D Blu-ray. 112 min. Original release 2013.

Captain America: The Winter Soldier (Marvel Studios/Walt Disney Studios Home Entertainment, 2014). 3D Blu-ray. 136 min. Original release 2014.

Guardians of the Galaxy (Marvel Studios/Walt Disney Studios Home Entertainment, 2014). 122 min. Original release 2014.

Avengers: Age of Ultron (Marvel Studios/Walt Disney Studios Home Entertainment, 2015). 3D Blu-ray. 141 min. Original release 2015.

Ant-Man (Marvel Studios/Walt Disney Studios Home Entertainment, 2015). 3D Blu-ray. 117 min. Original release 2015.

Captain America: Civil War (Marvel Studios/Walt Disney Studios Home Entertainment, 2016). 3D Blu-ray. 147 min. Original release 2016.

Doctor Strange (Marvel Studios/Walt Disney Studios Home Entertainment, 2016). 3D Blu-ray. 116 min. Original release 2016.

Guardians of the Galaxy, Vol. 2 (Marvel Studios/Walt Disney Studios Home Entertainment, 2017). 3D Blu-ray. 137 min. Original release 2017.

The Amazing Spider-Man (Marvel Studios/Sony Pictures Home Entertainment, 2012). 3D Blu-ray. 136 min. Original release 2012.

The Amazing Spider-Man 2 (Marvel Studios/Sony Pictures Home Entertainment, 2014). 3D Blu-ray. 142 min. Original release 2014.
Spider-Man: Homecoming (Marvel Studios/Sony Pictures Home Entertainment, 2017). 3D Blu-ray. 133 min. Original release 2017.
Thor: Ragnarok (Marvel Studios/Walt Disney Studios Home Entertainment, 2017). 3D Blu-ray. 130 min. Original release 2017.
Black Panther (Marvel Studios/Walt Disney Studios Home Entertainment, 2018). 3D Blu-ray. 134 min. Original release 2018.
Avengers: Infinity War (Marvel Studios/Walt Disney Studios Home Entertainment, 2018). 3D Blu-ray. 149 min. Original release 2018.
Ant-Man and the Wasp (Marvel Studios/Walt Disney Studios Home Entertainment, 2018). 3D Blu-ray. *118 min.* Original release 2018.
Captain Marvel (Marvel Studios/Walt Disney Studios Home Entertainment, 2019). 3D Blu-ray. 124 min. Original release 2019.
Avengers: Endgame (Marvel Studios/Walt Disney Studios Home Entertainment, 2019). 3D Blu-ray. 181 min. Original release 2019.
Spider-Man: Far from Home (Marvel Studios/Sony Pictures Home Entertainment, 2019). 129 min. Original release 2019.
Spider-Man: Into the Spider-Verse (Marvel Studios/Sony Pictures Home Entertainment, 2018). 3D Blu-ray. 117 min. Original release 2018.
Man of Steel (Warner Bros., 2013). 3D Blu-ray. 143 min. Original release 2013.
Batman v Superman: Dawn of Justice (Extended Cut; DC/Warner Bros., 2016). 3D Blu-ray Ultimate Edition. 182 min. Original release 2016.
Suicide Squad 3D (Extended Cut; DC/Warner Bros., 2016). 3D Blu-ray. 134 min. Original release 2016.
Wonder Woman (DC/Warner Bros., 2017). 3D Blu-ray. 141 min. Original release 2017.
Justice League 3D (DC/Warner Bros., 2017). 3D Blu-ray. 120 min. Original release 2017.
Aquaman 3D (DC/Warner Bros., 2019). 3D Blu-ray. 143 min. Original release 2018.
Shazam! (DC/Warner Bros., 2019). 3D Blu-ray. 132 min. Original release 2019.
Joker (Village Roadshow Pictures/Warner Bros. 2020). Blu-ray and 4K Ultra HD. 122 min. Original release 2019.
The Wolverine (Extended Cut; Marvel/20th Century Fox Home Entertainment, 2013). 3D Blu-ray Unleashed Extended Edition. 138 min. Original release 2013.
X-Men: Days of Future Past (Marvel/20th Century Fox Home Entertainment, 2014). 3D Blu-ray. 131 min. Original release 2014.
X-Men: Apocalypse (Marvel/20th Century Fox Home Entertainment, 2016) 3D Blu-ray. 144 min. Original release 2016.
Logan (Marvel/20th Century Fox Home Entertainment, 2017). Blu-ray and 4K Ultra HD. 137 min. Original release 2017.
Deadpool (Marvel/20th Century Fox Home Entertainment, 2016). Blu-ray. 108 min. Original release 2016.
Deadpool 2 (Marvel/20th Century Fox Home Entertainment, 2018). Blu-ray and 4K Ultra HD. 119 min. Original release 2018.
Dark Phoenix (Marvel/20th Century Fox Home Entertainment, 2019). Blu-ray and 4K Ultra HD. 114 min. Original release 2019.
Dolphin Tale (ALCON/Warner Bros. Home Video, 2011). 3D Blu-ray. 113 min. Original release 2011.

The Three Musketeers (Summit Entertainment/Constantin Film, 2011). 110 min. Original release 2011.

The Young and Prodigious T. S. Spivet (*El Extraordinario Viaje de T. S. Spivet* [Spanish version]; Europa Distribution, 2013). 3D Blu-ray. 105 min. Original release 2013.

Gravity (Warner Bros. Home Video, 2014). 3D Blu-ray. 91 min. Original release 2013.

The Walk (TriStar Pictures/Sony Pictures, 2015). 3D Blu-ray. 123 min. Original release 2015.

In the Heart of the Sea (Village Roadshow Pictures/Warner Home Video, 2016). 3D Blu-ray. Original release 2015.

Wolf Totem (Sony Pictures Home Entertainment, 2015). 3D Blu-ray. 121 min. Original release 2015.

Saw 3D: The Final Chapter (Lionsgate, 2010). 3D Blu-ray. 90 min. Original release 2010.

Resident Evil: Afterlife (Sony Pictures Home Entertainment, 2010). 3D Blu-ray. 97 min. Original release 2010.

Fright Night 3D (Touchstone Home Entertainment, 2011). 3D Blu-ray. 106 min. Original release 2011.

Final Destination 5 (New Line Cinema, 2011). 3D Blu-ray. 92 min. Original release 2011.

Paranormal Xperience 3D (Sony Pictures [Spain], 2014). 3D Blu-ray. Original release 2011.

Shark Night 3D (Rogue Pictures, 2011). 3D Blu-ray. 90 min. Original release 2011.

Piranha 3DD (Anchor Bay Entertainment, 2012). 3D Blu-ray. 82 min. Original release 2012.

Bait 3D (Anchor Bay Entertainment, 2012) 3D Blu-ray. 93 min. Original release 2012.

Dracula 3D (*Dracula di Dario Argento* [Italian version]; Sony Pictures Home Entertainment, 2013). 3D Blu-ray. 110 min. Original release 2012.

Abraham Lincoln: Vampire Hunter (20th Century Fox Home Entertainment, 2012). 3D Blu-ray. 105 min. Original release 2012.

Resident Evil: Retribution (Sony Pictures Home Entertainment, 2012). 3D Blu-ray. 96 min. Original release 2012.

Static (Flatiron Film Company, 2013). 3D Blu-ray. 90 min. Original release 2012.

Underworld: Awakening (Lakeshore/Sony Pictures Home Entertainment, 2012). 89 min. Original release 2012.

Tormented 3D (Well Go USA Entertainment [Japan], 2011. 3D Blu-ray. 83 min. Original release 2011.

Sadako 3D (Well Go USA Entertainment [Japan], 2012). 3D Blu-ray. 97 min. Original release 2012.

World War Z (Paramount Pictures, 2013). 3D Blu-ray New Unrated Cut. 122 min. Original release 2013.

Texas Chainsaw 3D (Lionsgate, 2013). 3D Blu-ray. 92 min. Original release 2013.

Spiders (Millennium Films, 2013). 3D Blu-ray. 89 min. Original release 2013.

Nurse 3D (Lionsgate, 2014). 3D Blu-ray. 84 min. Original release 2014.

Dracula Untold (Legendary Pictures/Universal Pictures, 2015). Blu-ray. 92 min. Original release 2014.

Poltergeist (MGM/20th Century Fox, 2015). 3D Blu-ray Extended Cut. 101 min. Original release 2015.

Paranormal Activity: The Ghost Dimension (Paramount Pictures, 2015). 3D Blu-ray New Unrated Cut. 95 min. Original release 2015.

The Mummy (Universal Studios, 2017). 3D Blu-ray. 110 min. Original release 2017.

The Meg (Warner Bros. Pictures, 2018). 3D Blu-ray. 113 min. Original release 2018.

Tron: Legacy (Walt Disney Studios Home Entertainment, 2010) 3D Blu-ray. 125 min. Original release 2010.

John Carter 3D (Walt Disney Studios Home Entertainment, 2012). 3D Blu-ray. 132 min. Original release 2012.

Upside Down (Millennium Entertainment, 2012). 3D Blu-ray. 107 min. Original release 2012.

Dredd (Lionsgate, 2013). 3D Blu-ray. 95 min. Original release 2012.

Men in Black 3 (Columbia Pictures/Sony Pictures Home Entertainment, 2012). 3D Blu-ray. 106 min. Original release 2012.

G. I. Joe: Retaliation (Paramount Pictures, 2013). 3D Blu-ray. 110 min. Original release 2013.

RoboCop (MGM/20th Century Fox Home Entertainment, 2014). Blu-ray. 118 min. Original release 2014.

I, Frankenstein 3D (Lionsgate, 2014). 3D Blu-ray. 92 min. Original release 2014.

Live. Die. Repeat /Edge of Tomorrow (Village Roadshow Pictures/Warner Bros. Pictures, 2014). 3D Blu-ray. 113 min. Original release 2014.

Independence Day: Resurgence (20th Century Fox Home Entertainment, 2016). 3D Blu-ray. 126 min. Original release 2015.

Terminator Genesys (Paramount Home Entertainment, 2015). 3D Blu-ray. 126 min. Original release 2015.

Jupiter Ascending (Warner Home Video, 2015). 3D Blu-ray. 127 min. Original release 2015.

Passengers (Columbia Pictures/Sony Pictures, 2016; also distributed in Oculus VR format). 3D Blu-ray and 4K Ultra HD. 116 min. Original release 2016.

Ready Player One (Warner Bros. Home Entertainment, 2018). 3D Blu-ray. 140 min. Original release 2018.

Mortal Engines (Universal Pictures Home Entertainment, 2018). 3D Blu-ray. 128 min. Original release 2018.

The Darkest Hour (Summit Entertainment, 2012). 3D Blu-ray. 89 min. Original release 2011.

Sector 7 (Shout Factory, 2011; South Korea). 3D Blu-ray. 104 min. Original release 2011.

San Andreas (Warner Bros. Home Entertainment, 2015). 3D Blu-ray. 114 min. Original release 2015.

Geostorm 3D (Warner Bros. Home Entertainment, 2017). 3D Blu-ray. 109 min. Original release 2017.

Skyscraper 3D (Universal Pictures Home Entertainment, 2018). 3D Blu-ray. Original release 2018.

Star Trek into Darkness 3D (Paramount Home Media Distribution, 2013). 3D Blu-ray. 131 min. Original release 2013.

Star Trek Beyond (Paramount Home Media Distribution, 2016). 3D Blu-ray. 122 min. Original release 2016.

Transformers: Dark of the Moon (Paramount Pictures, 2011). 3D Blu-ray Limited Edition. 154 min. Original release 2011.

Transformers: Age of Extinction (Paramount Pictures, 2014; Expanded images as seen in IMAX 3D theaters) 3D Blu-ray. 165 min. Original release 2014.

Transformers: The Last Knight (Paramount Pictures, 2017). 3D Blu-ray. 149 min. Original release 2017.

Dawn of the Planet of the Apes (20th Century Fox Home Entertainment, 2014). 3D Blu-ray. 130 min. Original release 2014.

War for the Planet of the Apes (20th Century Fox Home Entertainment, 2017). 3D Blu-ray. 140 min. Original release 2017.

Episode VII—The Force Awakens (Walt Disney Studios Home Entertainment, 2015). 3D Blu-ray Collector's Edition. 135 min. Original release 2015.

Episode VIII—The Last Jedi (Walt Disney Studios Home Entertainment, 2018). 3D Blu-ray. 152 min. Original release 2017.

Episode IX—The Rise of Skywalker (Walt Disney Studios Home Entertainment, 2020). 142 min. Original release 2019.

Rogue One: A Star Wars Story (Walt Disney Studios Home Entertainment, 2017). 3D Blu-ray. 135 min.

Jurassic World 3D (Legendary Pictures/Universal Pictures Home Entertainment, 2015). 3D Blu-ray. 124 min. Original release 2015.

Jurassic World: Fallen Kingdom (Legendary Pictures/Universal Picture Home Entertainment, 2018). 3D Blu-ray. 128 min. Original release 2018.

Pacific Rim 3D (Legendary Pictures/Universal Pictures Home Entertainment, 2013). 3D Blu-ray. 132 min. Original release 2013.

Pacific Rim: Uprising 3D (Legendary Pictures/Warner Bros. Pictures, 2018). 3D Blu-ray. 111 min. Original release 2018.

Godzilla (Legendary Pictures/Warner Bros. Pictures, 2014). 3D Blu-ray. 123 min. Original release 2014.

Kong: Skull Island (Legendary Pictures/Warner Bros. Pictures, 2017). 3D Blu-ray. 118 min. Original release 2017.

Godzilla: King of the Monsters (Legendary Pictures/Warner Bros. Pictures, 2019). 3D Blu-ray and 4K Ultra HD. 132 min. Original release 2019.

Stalingrad (Columbia Pictures/Sony Pictures, 2013). 3D Blu-ray. 131 min. Original release IMAX 3D, Russia, 2013.

Run 3D (Millennium Entertainment, 2013). 3D Blu-ray. 90 min. Original release 2013.

Everest (Universal Studios Home Entertainment, 2016). 3D Blu-ray. 121 min. Original release 2015.

Point Break (ALCON/Warner Bros. Pictures, 2015). 3D Blu-ray. 114 min. Original release 2015.

The Legend of Tarzan (Village Roadshow Pictures/Warner Bros. Pictures, 2016). 3D Blu-ray. 110 min. Original release 2016.

Mad Max: Fury Road (Warner Bros. Pictures, 2015). 3D Blu-ray. 120 min. Original release 2015.

Flying Swords of Dragon Gate (Vivendi Entertainment, 2011). 3D Blu-ray. 122 min. Original release 2011.

The Great Wall (Legendary Pictures/Universal Pictures Home Entertainment, 2016). 3D Blu-ray. 103 min. Original release 2016.

A Very Harold & Kumar 3D Christmas (New Line Cinema/Warner Bros. Pictures, 2012). 3D Blu-ray. 89 min. Original release 2011.

Spy Kids: All the Time in the World (Anchor Bay Entertainment, 2011). 3D Blu-ray. 88 min. Original release 2011.

R. I. P. D. 3D (Universal Studios Home Entertainment, 2013). 3D Blu-ray. 96 min. Original release 2013.

Pixels in 3D (Columbia Pictures/Sony Pictures, 2015). 3D Blu-ray. 106 min. Original release 2015.

Ghostbusters: Answer the Call (Columbia Pictures/Sony Pictures, 2016). 3D Blu-ray and 4K Ultra HD. 116 min. Original release 2016.

Ghost Rider: Spirit of Vengeance (Columbia Pictures/Sony Pictures, 2012). 3D Blu-ray. 95 min. Original release 2012.

Jack the Giant Slayer (New Line Cinema/Warner Bros. Pictures, 2013). 3D Blu-ray. 114 min. Original release 2013.
Oz the Great and Powerful (Walt Disney Studios Home Entertainment, 2013). 3D Blu-ray. Original release 2013.
Hansel & Gretel: Witch Hunters 3D (Paramount Home Media Distribution, 2013). 3D Blu-ray Unrated Cut. 97 min. Original release 2013.
Percy Jackson: Sea of Monsters (20th Century Fox Home Entertainment, 2013). 3D Blu-ray Deluxe Edition. 106 min. Original release 2013.
Pan (Warner Home Video, 2015). 3D Blu-ray. 111 min. Original release 2015.
The Huntsman: Winter's War (Universal Pictures Home Entertainment, 2016). 3D Blu-ray Extended Edition. 120 min. Original release 2016.
Miss Peregrin's Home for Peculiar Children (20th Century Fox Home Entertainment, 2016). 3D Blu-ray. 127 min. Original release 2017.
Pirates of the Caribbean: Dead Men Tell No Tales (*Pirates of the Caribbean: Salazar's Revenge* [UK version]; Walt Disney Studios Home Entertainment, 2017). 3D Blu-ray. 129 min. Original release 2017.
Harry Potter and the Deathly Hallows—Part 1 (Warner Bros. Pictures, 2011). 3D Blu-ray. 146 min. Original release 2010.
Harry Potter and the Deathly Hallows—Part 2 (Warner Bros. Pictures, 2011). 3D Blu-ray. 130 min. Original release 2011.
Fantastic Beasts and Where to Find Them (Warner Bros. Pictures, 2017). 3D Blu-ray. 133 min. Original release 2016.
Teenage Mutant Ninja Turtles 3D (Paramount Pictures, 2014). 3D Blu-ray. 101 min. Original release 2014.
The Divergent Series: Insurgent (Summit Entertainment, 2015). 3D Blu-ray. 119. Original release 2015.
The Hunger Games: Mockingjay—Part 1 (Lionsgate, 2015). Blu-ray. 123 min. Original release 2014.
The Hunger Games: Mockingjay—Part 2 (Lionsgate, 2016). Blu-ray. 137 min. Original release 2015.
Silent Hill: Revelation 3D (Universal Studios Home Entertainment, 2012). 3D Blu-ray. 95 min. Original release 2012. *Warcraft* (Duncan Jones, 2016).
Assassin's Creed (New Regency Productions, 2016). 3D Blu-ray. 115 min. Original release 2016.
Tomb Raider 3D (Warner Bros. Pictures, 2018). 3D Blu-ray 118 min. Original release 2018.
Allita: Battle Angel (20th Century Fox Home Entertainment, 2019). 3D Blu-ray. 122 min. Original release 2019.
Conan the Barbarian (Lionsgate Home Entertainment, 2011). 3D Blu-ray. 112 min. Original release 2011.
Immortals 3D (20th Century Fox Home Entertainment, 2012). 3D Blu-ray. 110 min. Original release 2011.
47 Ronin 3D (Universal Pictures Home Entertainment, 2014). 3D Blu-ray. 127 min. Original release 2013.
Pompeii (Constantin/Tristar/Sony Pictures, 2014). 3D Blu-ray. 114 min. Original release 2014.
Noah (Paramount Pictures, 2014; 3D abroad only). Blu-ray. 132 min. Original release 2014.
300: Rise of an Empire (Legendary Pictures/Warner Bros., 2014). 102 min. Original release 2014.

Hercules 3D (MGM/Paramount Pictures, 2014). 3D Blu-ray. 98 min. Original release 2014.

The Legend of Hercules (Millennium Films, 2014). 3D Blu-ray. Original release 2014.

Gods of Egypt (Summit Entertainment, 2016). 3D Blu-ray. 127 min. Original release 2016.

Ben-Hur (MGM/Paramount Pictures, 2016). Blu-ray. 125 min. Original release 2016.

King Arthur: Legend of the Sword (Village Roadshow Pictures/Warner Bros. Pictures, 2017). 3D Blu-ray. Original release 2017.

Sin City: A Dame to Kill For (Anchor Bay, 2014). 3D Blu-ray. 102 min. Original release 2014.

Murder on the Orient Express (20th Century Fox Home Entertainment, 2018). Blu-ray and 4K Ultra HD. 114 minutes. Original release 2017.

Cave of Forgotten Dreams (History Films/Sundance Selects, 2011). 3D Blu-ray. 89 min. Original release 2010.

Pina (Criterion Collection/Sundance Selects, 2013). 3D Blu-ray. 106 min. Original release 2011.

Goodbye to Language 3D (Kino Lorber, 2015). 3D Blu-ray. 69 min. Original release 2014.

Love 3D (Curzon Artificial Eye, 2015). 3D Blu-ray. Original release 2015.

Post-2009 Disney and/or Pixar Computer Animation

Tangled (Walt Disney Studios Home Entertainment, 2011). 3D Blu-ray. 100 min. Original release 2010.

Mars Needs Moms (Image Movers Digital/Buena Vista Home Entertainment, 2011). 3D Blu-ray. 88 min. Original release 2011.

Wreck-It Ralph (Walt Disney Studios Home Entertainment, 2013). 3D Blu-ray. 101 min. Original release 2012.

Frankenweenie (Walt Disney Pictures, 2013; stop-motion). 3D Blu-ray. 88 min. Original release 2012.

Finding Nemo (Pixar Animation Studios/Walt Disney Pictures; 3D conversion 2012). 3D Blu-ray. 100 min. Original release 2003.

Monsters, Inc. (Pixar Animation Studios/Walt Disney Pictures; 3D conversion 2012). 3D Blu-ray. 92 min. Original release 2001.

Brave (Mark Andrews and Brenda Chapman, 2012), *Monsters University* (Scanlon, 2013).

The Little Mermaid (Walt Disney Diamond Editions; 3D conversion, 2013). 3D Blu-ray. 83 min. Original release 1989.

Frozen (Walt Disney Studios Home Entertainment, 2014). 3D Blu-ray. 201 min. Original release 2013.

Ratatouille (Pixar Animation Studios/Walt Disney Pictures; 3D conversion 2014). 3D Blu-ray. 111 min. Original release 2007.

The Good Dinosaur (Pixar Animation Studios/Walt Disney Pictures, 2015). 94 min. Original release 2015.

Finding Dory (Walt Disney Studios Home Entertainment, 2016). 3D Blu-ray. 97 min. Original release 2016.

Zootopia (Walt Disney Studios Home Entertainment, 2016). 3D Blu-ray. 108 min. Original release 2016.

Moana (Walt Disney Studios Home Entertainment, 2017). 3D Blu-ray. Ultimate Collector's Edition. 103 min. Original release 2016.

Coco (Walt Disney Studios Home Entertainment, 2017). 3D Blu-ray. 105 min. Original release 2017. Unkrich, 2017.

Toy Story 4 (Pixar Animation Studios/Walt Disney Pictures, 2019). 3D Blu-ray and 4K Ultra HD. 100 min. Original release 2019. 104 min. 3D Blu-ray.

Frozen 2 (Walt Disney Animation, 2019). Blu-ray and 4K Ultra HD.

Disney live-action/ photorealistic CGI remakes of Disney animated classics

Alice in Wonderland 3D (Walt Disney Studios Home Entertainment, 2010). 3D Blu-ray. 108 min. Original release 2010, bases on the 1951 animated feature of the same title.

Maleficent (Walt Disney Studios Home Entertainment, 2014). 3D Blu-ray. 97 min. Original release 2014, based on the 1959 animated feature *Sleeping Beauty*.

Cinderella (Walt Disney Studios Home Entertainment, 2015). 3D Blu-ray. 116 min. Original release 2015, based on the 1950 animated feature of the same title.

The Jungle Book (Walt Disney Studios Home Entertainment, 2016). 3D Blu-ray. 106 min. Original release 2016, based on the 1967 animated feature of the same title.

Alice through the Looking Glass (Walt Disney Studios Home Entertainment, 2016). 3D Blu-ray. 113 min. Original release 2016, based on the 1951 animated feature *Alice in Wonderland*.

Beauty and the Beast (Walt Disney Studios Home Entertainment, 2017). 3D Blu-ray (UK version). 129 min. Original release 2017, based on the 1991 animated feature of the same title.

Christopher Robin (Walt Disney Studios Home Entertainment, 2018). Blu-ray. 104 min. Original release 2018, based on the 2011 animated feature *Winnie the Pooh* and numerous shorts/featurettes in the Disney Winnie-the-Pooh franchise.

Dumbo (Walt Disney Pictures, 2019). 3D Blu-ray (Sri Lankan version). 112 min. Original release 2019, based on the 1941 animated feature of the same title.

Aladdin (Walt Disney Studios Home Entertainment, 2019). Blu-ray and 4K Ultra HD. 128 min. Original release 2019, based on the 1992 animated feature of the same title.

The Lion King (Walt Disney Studios Home Entertainment, 2019). Blu-ray and 4K Ultra HD. Original release 2019, based on the animated feature of the same title.

Maleficent: Mistress of Evil (Walt Disney Studios Home Entertainment, 2019). Blu-ray and 4K Ultra. Original release 2019, based on the 1959 animated feature Sleeping Beauty.

Lady and the Tramp 104 min. (Charlie Bean, 2019; original was 1955).

Mulan (Niki Caro, 2020; original was 1998).

For the nearly 250 computer-animated, non-Walt Disney Pictures and/or Pixar Animation Studios digital 3D titles, the reader is referred to the list of "**3D FEATURES PRODUCED 2010-20,** ANIMATION (MAINLY 3D COMPUTER)" noted in the text. Nearly all of those features are available online from the distribution companies under which they are listed there.

This Discography shows two trends. First, as the popularity of 3D grew during the 2010s, the window between the theatrical and home theater release of successful films grew shorter and shorter, until near the end of the decade they were nearly simultaneous. (In fact, in some cases, such as that of *Prometheus* in 2012, the home media release *preceded* the theatrical release by several weeks.) Second, after 2017, 3D Blu-ray release was increasingly accompanied or by replaced by 4K Ultra HD to conform to the reconfigured home media market. While some potential blockbusters (e.g., *Frozen 2* [2019], *Star Wars: The Rise of Skywalker* [2019]) continued to be released in 3D versions theatrically, their home media release in Western markets was invariably in standard Blu-ray and/or 4K Ultra HD format.

In and after 2020, there were only a handful of 3D Blu-ray releases, and these were mainly archival in nature.

SELECTED BIBLIOGRAPHY

Abel, Richard (ed.). *The Encyclopedia of Early Cinema*. London: Routledge, 2005.

Adler, Dan, Janine Marchessault, and Sanja Obrodovic (eds.). *3D Cinema and Beyond*. London: Intellect, 2014.

Atkinson, Sarah. *Beyond the Screen: Emerging Cinema and Engaging Audiences*. London: Bloomsbury Academic Press, 2015; rpt.

Bailenson, Jeremy. *Experience on Demand: What Virtual Reality Is, How It Works, and What It Can Do*. New York: W. W. Norton, 2018.

Balzer, Richard. Peepshows: *A Visual History*. New York: Harry N. Abrams, 1998.

Barker, Jennifer M. *The Tactile Eye: Touch and the Cinematic Experience*. Berkeley: University of California Press, 2009.

Bazin, André. *What Is Cinema?*, 2 volumes, trans. Hugh Gray. Berkeley: University of California Press, 1967.

Block, Bruce, and Philip McNally. *3D Storytelling: How Stereoscopic 3D Works and How to Use It*. New York: Focal Press, 2013.

Brownlow, Kevin. *The Parade's Gone By*. New York: Bonanza Books, 1968.

Colin, Armand. *Le cinema 3-D—Histoire, economie, technique, esthetique*. Paris: Hachette Book Group, 2015.

Cook, David A. *A History of Narrative Film*, fourth edition. New York: W. W. Norton, 2004.

———. *A History of Narrative Film*, fifth edition. New York: W. W. Norton, 2016.

———. *Lost Illusions: American Cinema in the Shadow of Watergate and Vietnam*. New York: Scribner's, 2000.

Cook, Olive. *Movement in Two Dimensions: A Study of the Animated and Projected Pictures That Preceded the Invention of Cinematography*. London: Hutchinson, 1963.

Coppola, Francis Ford. *Live Cinema and Its Techniques*. New York: W. W. Norton, 2017.

Crary, Jonathan. *Techniques of the Observer: On Vision and Modernity in the Nineteenth Century*. Cambridge, MA: MIT Press, 1990.

Delson, Susan. "The Many Sides of 3-D: From 1830s Science to 1950s Monster Movies, from Avant-Garde Experimenters to Recent Hits: A Survey in L. A." *Wall Street Journal* (June 30–July 1, 2018).

Dsouza, Clyde. *Think in 3D: Food for Thought for Directors, Cinematographers, and Stereographers*. Los Angeles: Create Space Independent Publishing Platform, 2012.

Edward, Graham. "The Dreamsmiths Unleashed." *Cinefex* 151 (February 2017): 79–98.

Elsaesser, Thomas. *The Persistence of Hollywood*. London: Routledge, 2012.

Frizot, Michel. *A New History of Photography*. London: Konemann, 1998.

Gelernter, David. "Machines That Will Think and Feel." *Wall Street Journal* (March 18, 2016).

Grau, Oliver. *Virtual Art: From Illusion to Immersion*, trans. Gloria Custance. Cambridge, MA: MIT Press, 2003.

Griffiths, Alison. *Shivers Down Your Spine: Cinema, Museums and the Immersive View*. New York: Columbia University Press, 2008.

Hayes, R. M. *3-D Movies: A History and Filmography of Stereoscopic Cinema*, revised edition. Jefferson, NC: McFarland, 1998.

Herbert, Stephen, and Luke McKernan (eds.). *Who's Who of Victorian Cinema*. London: BFI, 1996.

Huhtamo, Erkki. *Illusions in Motion: Media Archaeology of the Moving Panorama and Related Spectacles*. Cambridge, MA: MIT Press, 2013.

Isaacs, Bruce. *The Orientation of Future Cinema: Technology, Aesthetics, Spectacle*. London: Bloomsbury Academic, 2012.

Jabr, Ferris. "Can You Really Be Addicted to Video Games?" *New York Times* (October 27, 2019): 36–41; 54–55.

Kelly, Kevin. "Hypervision," *Wired* 24, 5 (May 2016): 74–87; 112.

Klinger, Barbara. *Beyond the Multiplex: Cinema, New Technologies, and the Home*. Berkeley: University of California Press, 2006.

Kroon, Richard W. *3D A-to-Z: An Encyclopedic Dictionary*. Jefferson, NC: McFarland, 2012.

Leeman, Fred. *Hidden Images: Games of Perception, Anamorphic Art, Illusion from the Renaissance to the Present*, trans. Ellyn Childs Allison and Mike Schuyt. New York: Harry N. Abrams, 1976.

Leslie, Esther. *Hollywood Flatlands: Animation, Critical Theory and the Avant-garde*. London: Verso, 2002.

Lewis, Jon. *Hollywood V. Hardcore: How the Struggle over Censorship Created the Modern Film Industry*. New York: New York University Press, 2002.

Limbacher, James L. *Four Aspects of the Film: 3-D, Color, Sound, 3-D, Widescreen*. New York: Brussel and Brussel, 1968.

Lippet, Akira Mizuta. *Atomic Light (Shadow Optics)*. Minneapolis: University of Minnesota Press, 2005.

Lipton, Lenny. *Foundations of the Stereoscopic Cinema: A Study in Depth*. New York: Van Nostrand Reinhold, 1982.

Manovich, Lev. *The Language of New Media*. Cambridge, MA: MIT Press, 2002.

———. *Software Takes Command*. London: Bloomsbury Academic Books, 2013, 376.

Marks, Laura U. *The Skin of the Film: Intercultural Cinema, Embodiment, and the Senses*. Durham, NC: Duke University Press, 2000.

Mendiburu, Bernard. *3D Movie Making: Stereoscopic Digital Cinema from Script to Screen*. New York: Focal Press, 2009.

———. *3D TV and 3D Cinema: Tools and Processes for Creative Stereography*. New York: Focal Press, 2011.

Michel, Bernoit J., and Colette Michel. *Digital Stereoscopy: Scene to Screen 3D Production Workflow*. Los Angeles: Create Space Independent Publishing Platform, 2013.

Mochizuki, Takeshi. "Videogame Business Has Its Head in the Cloud." *Wall Street Journal* (October 25, 2019): R10.

Morgan, Hal, and Dan Stymes. *Amazing 3-D*. Boston, MA: Little, Brown, 1982.

Musser, Charles. *The Emergence of Cinema: The American Screen to 1907*, Volume 1 of the History of American Cinema Series. New York: Charles Scribner's, 1990.

———. *Before the Nickelodeon: Edwin S. Porter and the Edison Manufacturing Company*. Berkeley: University of California Press, 1991.

Nazareiff, Serge. *The Stereoscopic Nude, 1850–1930*. Berlin: Benedikt Taschen Verlag Berlin Gmbtl, 1990.

Norling, J. A. "Three Dimensional Motion Pictures." *Society of Motion Picture Engineers Journal* (December 1939): 612.

Paul, William. *When Movies Were Theater: Architecture, Exhibition, and the Evolution of American Film*. Columbia University Press, 2016.

Pellerin, Denis, and Brian May. *The Poor Man's Picture Gallery: Stereoscopy versus Painting in the Victorian Era*. London: London Stereoscopic Company, 2014.

Pellerin, Denis, Brian May, and Paula Fleming. *Diableries: Stereoscopic Adventures in Hell*. London: London Stereoscopic, 2013.

Pennington, Adrian, and Carolyn Giardina. *Exploring 3D: The New Grammar of Stereoscopic Filmmaking*. New York: Focal Press, 2012.

Perkowitz, Sidney. *Empire of Light: A History of Discovery in Science and Art*. New York: Henry Holt, 1996.

Prince, Stephen. *Digital Effects in Cinema: The Seduction of Reality*. Brunswick, NJ: Rutgers University Press, 2012.

———. *Digital Cinema*. Brunswick, NJ: Rutgers University Press, 2019.

Purse, Lisa. *Digital Imaging in Popular Cinema*. Edinburgh: Edinburgh University Press, 2013.

Rogers, Ariel. *Cinematic Appeals: The Experience of New Movie Technologies*. New York: Columbia University Press, 2013.

Rombes, Nicholas. *Cinema in the Digital Age*, rev. ed. London: Wallflower Press, 2017.

Ross, Miriam. *3D Cinema: Optical Illusions and Tactile Experiences*. London: Palgrave Macmillan, 2015.

Rubin, Peter. *Future Presence: How Virtual Reality Is Changing Human Connection, Intimacy, and the Limits of Ordinary Life*. New York: HarperCollins, 2018.

Schroter, Jens. *3D: History, Theory and Aesthetics of the Transplane Image*, rev. ed., trans. Brigitte Pichon and Dorian Rudnytsky. London: Bloomsbury Academic, 2014.

Schwartzel, Eric. "Virtual Reality Movies: Get Ready for the VR Revolution." *Wall Street Journal* (March 4, 2016).

Siapera, Eugenia. *Understanding New Media*, 2nd ed. Thousand Oaks, CA: SAGE, 2018.

Silverman, Kaja. *The Miracle of Analogy: Or the History of Photography, Part 1*. Stanford, CA: Stanford University Press, 2015.

Smith, Michael D., Peter Lude, and Bill Hogan, eds. *3D Cinema and Television Technology: The First 100 Years: Selected Papers from the Society of Motion Picture and Television Engineers*. SMPTE, 2011.

Sobchak, Vivian. *Carnal Thoughts: Embodiment and Moving Image Culture*. University of California Press, 2004.

Tricart, Celine. *3D Filmmaking: Techniques and Best Practices for Stereoscopic Filmmakers*. New York: Focal Press, 2016.

———. *Virtual Reality Filmmaking: Techniques and Best Practices for VR Filmmakers*. New York: Focal Press, 2018.

Verhoeff, Nana. *Mobile Screens: The Visual Regime of Navigation*. Amsterdam: Amsterdam University Press.

Weetch, Owen. *Expressive Spaces in Digital 3D Cinema*. London: Palgrave Macmillan, 2016.

Williams, Linda. *Hard Core: Power, Pleasure, and the "Frenzy of the Visible."* Berkeley: University of California Press, 1989.

Wysotsky, Michael Z. *Wide-Screen Cinema and Stereophonic Sound*, trans. A. E. C. York. London: Focal Press, 1971.

Zacks, Jeremy M. *Flicker: Your Brain on Movies*. New York: Oxford University Press, 2015.

Zone, Ray, ed. *3D Filmmakers: Conversations with Creators of Stereoscopic Pictures*. Metuchen, NJ: Scarecrow Press, 2005.

Zone, Ray. *Stereoscopic Cinema and the Origins of 3-D Film, 1838–1952*. Lexington: University of Kentucky Press, 2007.

———. *3-D Revolution: The History of Modern Stereoscopic Cinema*. Lexington: University of Kentucky Press, 2012.

http://www.stereoscopicmedia.org.

http://www.3dfilmarchive.com

Over the past decade, I have accumulated a large collection of 3-D Blu-ray discs representing much of the history of stereoscopic cinema from the 1920s through the present. These titles are included in the extensive Discography above.

INDEX

www.ingramcontent.com/pod-product-compliance
Ingram Content Group UK Ltd.
Pitfield, Milton Keynes, MK11 3LW, UK
UKHW041831100926
8556IPUK00002B/118